ENGLISH FOR ACADEMIC PURPOSES SERIES

General Editor: Vaughan James

AGRICULTURE

Christopher St J Yates

ENGLISH LANGUAGE TEACHING

Prentice Hall

New York London Toronto Sydney Tokyo Singapore

Published 1992 by
Prentice Hall International (UK) Limited
Campus 400, Maylands Avenue
Hemel Hempstead, Herts HP2 7EZ
a division of
Simon & Schuster International Group

First published 1989 by Cassell Publishers Limited.

Printed and bound in Great Britain by The Bath Press, Avon.

A series designed and developed by Passim Ltd, Oxford, and Associates

British Library Cataloguing in Publication Data

Information available from the publisher on request.

ISBN 0-13-280132-6

1 2 3 4 5 96 95 94 93 92

CONTENTS

ACKNOWLEDGEMENTS

The Editor and Publishers would like to thank the following for permission to reproduce copyright material in the text: Buckett, *Introduction to Livestock Husbandry*, Pergamon Press, 1980; Barden, Halfacre & Parrish, *Plant Science*, McGraw-Hill, 1987; Janick *et al.*, *Plant Science*, W.H. Freeman, 1980; Joy & Wibberley, *A Tropical Agriculture Handbook*, Cassell, 1979; Kay, *Farm Management*, 2nd ed., McGraw-Hill, 1986; Lockhart & Wiseman, *Introduction to Crop Husbandry*, 6th ed., Pergamon Press, 1988; Shippen, Ellin & Clover, *Basic Farm Machinery*, 3rd ed., Pergamon Press, 1980.

Every reasonable effort has been made to contact copyright holders. The publishers will make appropriate arrangements with any such body with whom contact was not possible.

INTRODUCTION

This course has three purposes. It is intended:

- to introduce you to the **contents** of Agriculture Science:
- to provide examples of authentic texts written in the **language** typical of the subject:
- to help you to practise the **skills** you will need in order to study the subject via English and to use it when you have learned it.

No knowledge of Agriculture Science is assumed, but if you work through the book carefully you will certainly learn a great deal about it. We don't set out to give comprehensive coverage, but the material does embrace most of the basic concepts. In this sense it is a basic textbook of Agriculture Science.

All the texts are taken from publications about Agriculture Science. They are not simplified for learners of English: the language you will encounter in them is exactly what you will meet in real life. We assume that you will have already taken a course of general English and are familiar with the main grammatical structures and much of the vocabulary of everyday use. There may be no such thing as Agriculture Science English, but there are a number of words and expressions commonly used in Agriculture Science contexts and there are a number of structures also in common use, and these have been isolated for you to practise. So in this sense, this is a textbook of English.

The most important aim of the course, however, is to help you to acquire and develop the skills you will need in order to learn your subject and, when you have finished the course, to use what you will have learned.

When you begin to study a new subject, you do so in two main ways: by **reading** and by **listening**. These are the major means of access to new knowledge and it is on these that we concentrate, via the **book** — for reading, and the **tapes** — for listening. In order to attack all these aims, we have divided each of the 15 units into 8 sections, closely related but each with a slightly different emphasis. Below we give a brief description of each section, so that at any point in any unit you will know exactly what you are expected to do and why you are doing it. The pattern is the same for all units.

A. UNDERSTANDING A PRINTED TEXT (1): In this section you are given a passage to read, usually including a diagram or table, to introduce the topic of the unit. You should first read it through, even if you don't understand it all, looking especially at the way it is set out in paragraphs, with side headings, marginal notes, captions, etc. This will give you a general idea of what it is about and how it is arranged. You will probably need to read it several times.

B. CHECK YOUR UNDERSTANDING: To help you to identify the most important points in the reading passage, a small number of questions are given, the answers to which you can look out for as you read. You could tackle them by jotting down a few notes and then turning your notes into complete answers, which your teacher will check. When you are clear about the general meaning of the passage, you can work through it in more detail with your dictionary. You should *always* have a dictionary handy and *never* be too proud (or too lazy!) to look things up.

C. INCREASE YOUR VOCABULARY: In this section you are asked to look at certain words which are used in the text, and there are several kinds of activity to help you remember them. Notice that they are not all new or technical terms; it is often familiar words used in an unfamiliar way that will cause you trouble.

D. CHECK YOUR GRAMMAR: There are probably no new grammatical structures in the texts, but you may need reminding of some of them. The most important ones arising from the texts are revised and practised in this section.

E. UNDERSTANDING A LECTURE / H. UNDERSTANDING DISCOURSE: Sections A–D are all concerned with gaining access to new information through reading, but an important source of information is through listening — to lectures, talks, discussions, even simple conversations between fellow students — so sections E and H are both based on the recordings, to which you should listen (usually several times) before attempting to answer the questions or perform the activities given in your book. You will hear a variety of voices and accents, all speaking at the sort of speed that is customary in an English-speaking environment.

F. UNDERSTANDING A PRINTED TEXT (2) / G. CHECK YOUR UNDERSTANDING: These two sections are very similar to A and B, but the questions in section G are far more detailed and you will need to study the text very carefully in order to answer them.

Although we hope that you will enjoy working through this course, we do not expect you to find it easy. At various times you will probably start wondering how much you have been learning — or your teacher will want to find out what progress you are making. So after Units 5 and 10 we have included progress checks (not tests!) so that you can get a fairly clear idea of this. By the time you have completed Unit 15, you will be ready for anything!

Some of the texts are written in American English, which has some differences from British English, especially in spelling (e.g. *Br.* vapour, *Am.* vapor). You might find it useful to keep a running list of such examples. Remember that they are equally acceptable, but you should avoid mixing them in a single piece of writing.

Vaughan James

Oxford, 1989

A. Understanding a printed text (1)

The following text will introduce you to the topic of **plants**. Look at the way it is divided into sections and paragraphs. Pay attention to the headings and notes in the margins, and to the illustrations and captions.

Now look at these questions:

1. The text has three sections. What is each section about?
2. What kind of words are in *italics*?
3. What is the purpose of the chemical formula?
4. What section does Figure 1 relate to?
5. What does Figure 3 illustrate?

Read the passage through and find the answers to the questions. Remember, you do not have to understand every word to answer the questions.

Plants

What they are

1 PLANTS are living organisms consisting of innumerable tiny cells. They differ from animals in many ways but the most important difference is that plants can build up valuable organic substances from simple materials. The most important part of this building process, which is called *photosynthesis*, is the production of *carbohydrates* such as *sugars*, *starches* and *cellulose*.

Photosynthesis

The process

2 In photosynthesis a special green substance called *chlorophyll* uses *light* energy (normally sunlight) to change *carbon dioxide* and *water* into *sugars* (carbohydrates) in the *green* parts of the plant. The daily amount of photosynthesis is limited by the duration and intensity of sunlight. The amount of carbon dioxide available is also a limiting factor. Shortage of water and low temperatures can also reduce photosynthesis.

3 The cells which contain chlorophyll also have yellow pigments such as *carotene*. Crop plants can only build up chlorophyll in the light and so any leaves which develop in the dark are yellow and cannot produce carbohydrates.

4 *Oxygen* is released during photosynthesis and the process may be set out as follows:

$$\text{Carbon dioxide} + \text{water} + \text{energy} \xrightarrow{\text{chlorophyll}} \text{carbohydrates} + \text{oxygen}$$

$$nCO_2 \quad nH_2O \quad \text{(light)} \quad = \quad (CH_2O)n \quad nO_2$$

5 This process not only provides the basis for all our food but it also supplies the oxygen which animals and plants need for respiration.

Carbohydrates

6 The simple carbohydrates, such as *glucose*, may build up to form *starch* for storage purposes, or to *cellulose* for building cell walls. *Fats* and *oils* are formed from carbohydrates. *Protein* material, which is an essential part of all living cells, is made from carbohydrates and nitrogen compounds.

The parts of a plant

7 Most plants consist of *roots, stems, leaves* and *reproductive parts* and need *soil* in which to grow.

8 The *roots* spread through the spaces between the particles in the soil and anchor the plant. In a plant such as wheat the root system may total many miles.

9 The *leaves*, with their broad surfaces, are the main parts of the plant where photosynthesis occurs (see Fig. 1).

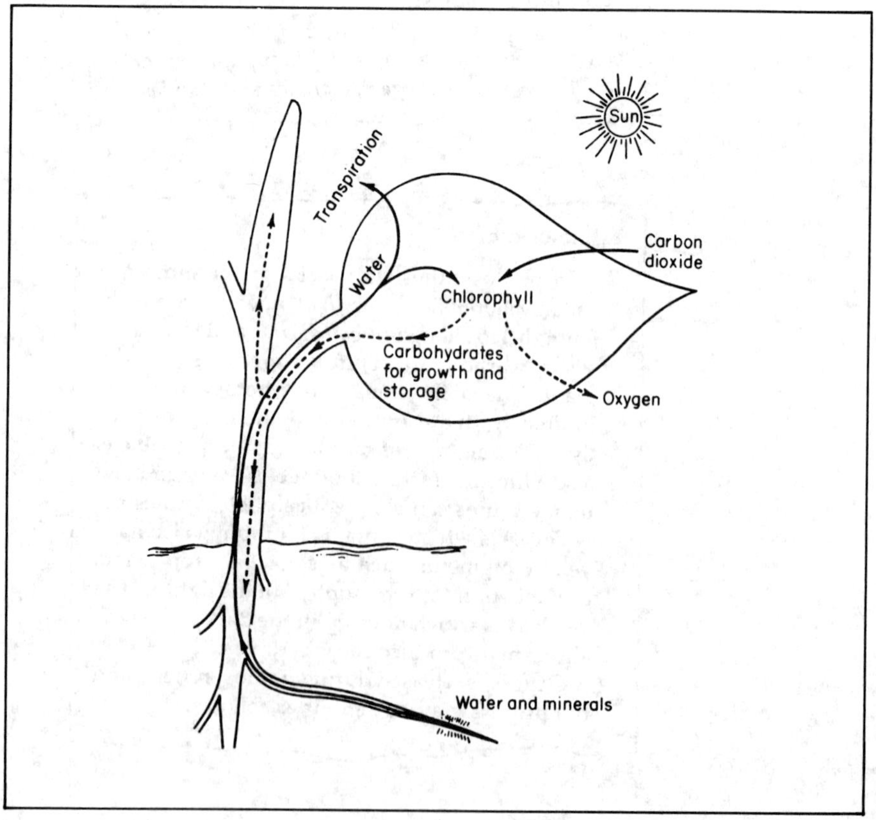

Figure 1 Photosynthesis illustrated diagrammatically.

10 A very important feature of the leaf structure is the presence of large numbers of tiny pores (*stomata*) on the surface of the leaf (see Fig. 2). There are usually thousands of stomata per square cm of leaf surface. Each pore (stoma) is oval-shaped and surrounded by two guard cells. When the guard cells are turgid (full of water) the stoma is open and when they lose water the stoma closes.

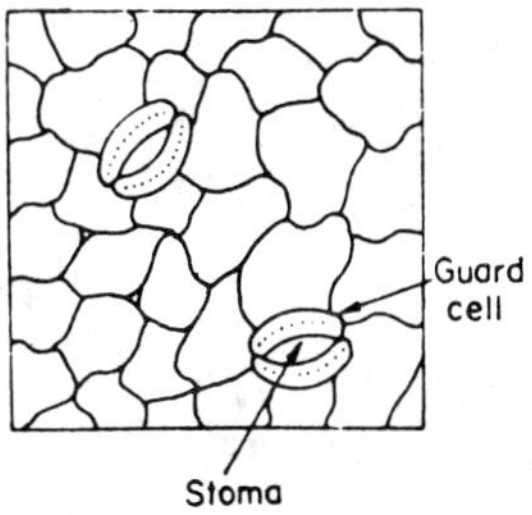

Figure 2 Stomata on leaf surface.

11 The carbon dioxide used in photosynthesis *diffuses* into the leaf through the stomata and most of the water vapour leaving the plant, and the oxygen from photosynthesis diffuses out through the stomata.

Transpiration

Evaporation of water

12 The evaporation of water from plants is called *transpiration*. It mainly occurs through the stomata and has a cooling effect on the leaf cells. Water in the cells of the leaf can pass into the pore spaces in the leaf and then out through the stomata as water vapour (see Fig. 3).

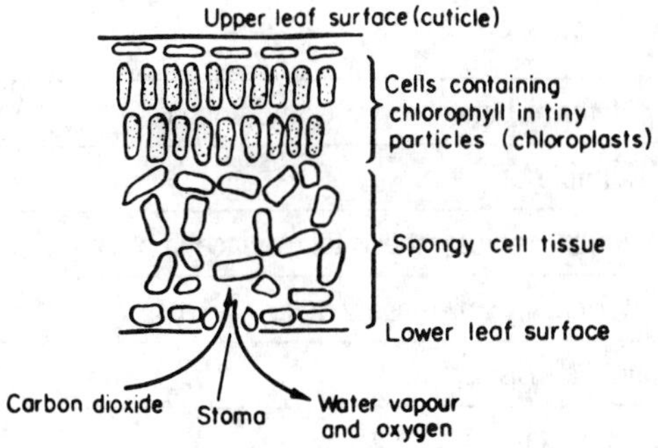

Figure 3 Cross-section of green leaf showing gaseous movements during daylight.

13 The rate of transpiration varies considerably. It is greatest when the plant is well supplied with water and the air outside the leaf is warm and dry. In very hot or windy weather water evaporates from the guard cells and so the stomata close and reduce the rate of transpiration. The stomata also close in very cold weather, e.g. 0°C.

14 The rate of loss is reduced if the plant is short of water because the guard cells then lose water and close the stomata; it is also retarded if the humidity of the atmosphere is high.

15 The stomata guard cells close (and so transpiration ceases) during darkness. They close because photosynthesis ceases and water is lost from the guard cells (osmosis) when some of the sugars present change to starch.

This text was taken (edited) from LOCKHART & WISEMAN, *Introduction to Crop Husbandry*, 5th edn., Pergamon Press

B. Check your understanding

Now read the text carefully, looking up any new items in a dictionary or reference book. Then answer the following questions:

1. What is the main difference between plants and animals?
2. What is starch an example of?
3. What decides how much photosynthesis takes place in a day?
4. What is the reason for crop plants having yellow leaves?
5. What do plants mainly consist of?
6. Where do you find stomata?
7. How does oxygen come out of a plant?
8. What else comes out of a plant?
9. Why do stomata close in hot weather?
10. Why does photosynthesis not take place in darkness?

C. Increase your vocabulary

In this section you should use your dictionary to help you answer the questions about the text.

1. Look at the first paragraph and say what these words refer to:
- line 2: They
- line 6: which

2. Look at paragraph 1 again and say which words in the text have the same meaning as:
- very small
- cannot be counted
- not complicated

3. Look at paragraph 2 and say what words have the opposite meaning to:
- cannot be obtained
- increase
- excess

4. Look at paragraph 2 again and say which words have the same meaning as:
- length of time
- alter
- quantity

5. Look at paragraphs 3, 4 and 5 and explain the following words:
- chlorophyll
- carotene
- respiration

6. Look at paragraph 6 again. Which words correspond to the definitions?
- a tiny unit of living matter
- something made up of two or more combined parts
- a body-building substance which is important for good health
- necessary

7. Look at paragraphs 7, 8 and 9 and say which words have the opposite meaning to:
- narrow
- does not happen

8. Look at paragraphs 10 and 11 and explain the following words:
- structure
- surface
- diffuses
- vapour

D. Check your grammar

MAKING STATEMENTS

> **Do you remember?**
> Plants *take in* oxygen which *combines* with organic foodstuffs.
> Crops *are checked* in growth if the roots *are deprived* of oxygen.

1. Use the following verbs to complete the paragraphs below: carry, be, help, perforate, carry, be, not lignify, usually find, form, thicken, make, know

The movement of materials through the plant _____ as *translocation*. The *xylem* or *wood vessels* which _____ the water and mineral salts from the roots to the leaves _____ tubes. These tubes _____ from dead cells. The cross walls of the cells have disappeared and the longitudinal walls _____ with *lignum*. These _____ wood. The tubes _____ to strengthen the stem. The *phloem tubes (bast)* _____ organic material through the plant. These vessels _____ chains of living cells which _____. They have cross walls which _____ – hence the altnerative name, *sieve tubes.* In the stem the xylem and phloem tubes _____ in a ring near the outside of the stem.

▼

Do you remember?
John grew barley and wheat last year.
Rice is eaten in many parts of the world.

2. Now write the following sentences in full, like this:
love apples/ago/many/years/as/tomatoes . . . . (know)
Many years ago tomatoes were known as love apples.

- manufacturer/by/this/the/but/week/it/a . . . . (repair, take)
- a/Peter/glasshouse/year/large/last . . . . (buy)
- successfully/and/they . . . . (germinate, prick out)
- crop/his/blight/finally/by . . . . (affect)
- trays/seeds/in/the/time/on . . . . (sow)
- of/second/plants/attack/this/the/many . . . . (distort)
- first/profit/he/a lot of/crop/his/on . . . . (not make)
- it/he/he/tomatoes/to grow/early/because . . . . (buy, want)
- plants/aphids/weeks/young/by/later/the/three . . . . (attack)
- crop/the/more/Peter/his/than/cost/he/but . . . . (save, be, expect)
- disaster/when/first/sprinkler system/the/the/then/but . . . . (strike, fail)
- then/the/plants/again/but . . . . (spray, attack)
- prepared/them/later/they/in/beds/the/for . . . . (plant out)

3. Now arrange the sentences you have made into a single paragraph. Make sure that the order you arrange the sentences in makes sense!

E. Understanding a lecture

1. You are now going to hear part of a lecture, divided into short sections to help you understand it. As you listen, answer the following questions:

Section 1

- What is the title of the lecture?
- Is this statement correct or incorrect?
  Plants get water from the soil. ☐
- Is this statement correct or incorrect?
  Plants absorb water through their root hairs. ☐
- The lecturer does not mention one term shown in the diagram. Which is it?

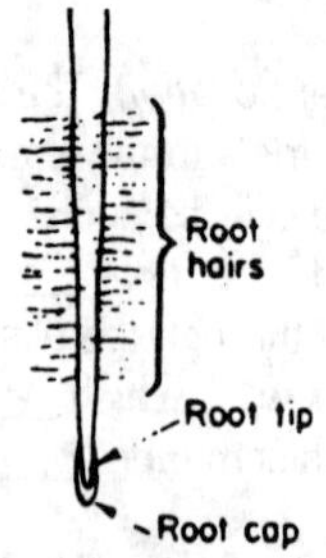

Section of root tip and root hair region

Section 2

- Complete the following statement correctly:
  The suction pull starts in a plant's
  root hairs ☐
  leaves ☐
  xylem tubes ☐
  root cells ☐
- The lecturer asks if you remember a particular term. What is it?
- How does the lecturer say water is absorbed into the roots?

Section 3

- Is this statement correct or incorrect?
  A plant takes in a constant amount of water throughout its life. ☐
- Note down the first reason why the rate of absorption can be slowed down.

Section 4

- Note down the second reason.
- What does the term *waterlogged* mean?

Section 5

- Write down the word the lecturer spells for you.
- Note down the third reason for a slow absorption rate.

2. Now wind the tape back to the beginning of the lecture and listen to it again. This time, instead of answering questions, take notes. The questions you have already answered will help you do this. When you have listened to the whole of the lecture, you will be asked to provide a short summary of it, using your notes as a guide. So make sure you note down the most important points.

F. Understanding a printed text (2)

Read the following text carefully, looking up anything you do not understand.

Plant Groups

1 Plants can be divided into annuals, biennials and perennials according to their total length of life.

Annuals

2 Typical examples are wheat, barley and oats which complete their life history in one growing season, i.e. starting from the seed, in 1 year they develop roots, stem and leaves and then produce flowers and seed before dying.

Biennials

3 These plants grow for 2 years. They spend the first year in producing roots, stem and leaves, and the following year in producing the flowering stem and seeds, after which they die.

4 Sugar beet, swedes and turnips are typical biennials, although the grower treats these crops as annuals, harvesting them at the end of the first year when all the foodstuff is stored up in the root.

Perennials

5 They live for more than 2 years and, once fully developed, they usually produce seeds each year. Many of the grasses and legumes are perennials.

This text was taken (edited) from LOCKHART & WISEMAN, *Introduction to Crop Husbandry*, 5th edn., Pergamon Press

Structure of the seed

6 Plants are also classified as *dicotyledons* and *monocotyledons* according to the structure of the seed.

Dicotyledon

7 A good example of a dicotyledon seed is the broad bean because it is large and easy to study. If a pod of the broad bean plant is opened when it is nearly ripe it will be seen that each seed is attached to the inside of the pod by a short stalk called the *funicle*. All the nourishment which the developing seed requires passes through the funicle from the bean plant.

8 When the seed is ripe and has separated from the pod a black scar, known as the *hilum*, can be seen where the funicle was attached. Near one end of this hilum is a minute hole called the *micropyle* (see Fig. 5).

9 If a bean is soaked in water the seed coat can be removed easily and all that is left is largely made up of the *embryo (germ)*. This consists of two seed leaves, or *cotyledons*, which contain the food for the young seedling.

10 Lying between the two cotyledons is the *radicle*, which eventually forms the *primary root*, and a continuation of the radicle the other end, the *plumule* (see Fig. 6). This develops into the young *shoot*, and is the first *bud* of the plant.

Monocotyledon

11 This class includes all the cereals and grasses and it is, therefore, very important.

12 The wheat grain is a typical example. It is not a true seed (it should be called a single-seeded fruit). The seed completely fills the whole grain, being practically united with the inside wall of the grain or fruit.

13 This fruit wall is made up of many different layers which are separated on milling into varying degrees of fineness, e.g. bran and pollards, and these are valuable livestock feed.

14 Most of the interior of the grain is taken up by the floury *endosperm*. The embryo occupies the small raised area at the base. The *scutellum*, a shield-like structure, separates the embryo from the endosperm. Attached to the base of the scutellum are the five roots of the embryo, one primary and two pairs of *secondary* rootlets. The roots are enclosed by a sheath called the *coleorhiza*. The position of the radicle and the plumule can be seen in the diagram (Fig. 7).

15 The scutellum can be regarded as the cotyledon of the seed. There is only *one* cotyledon present and so wheat is a monocotyledon.

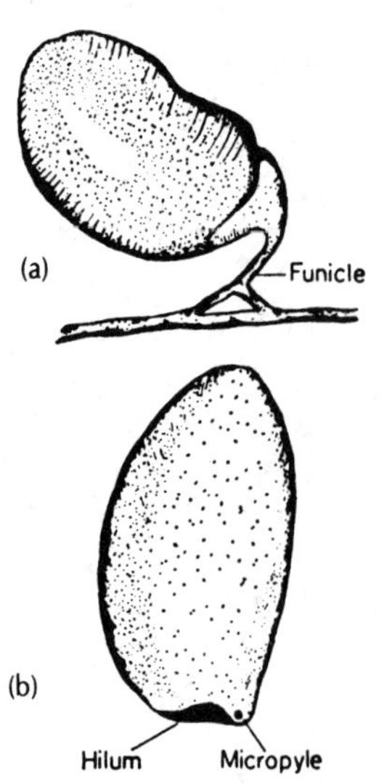

Figure 5 (a) Bean seed attached to the inside of the pod by the funicle, (b) bean seed showing the hilum and micropyle.

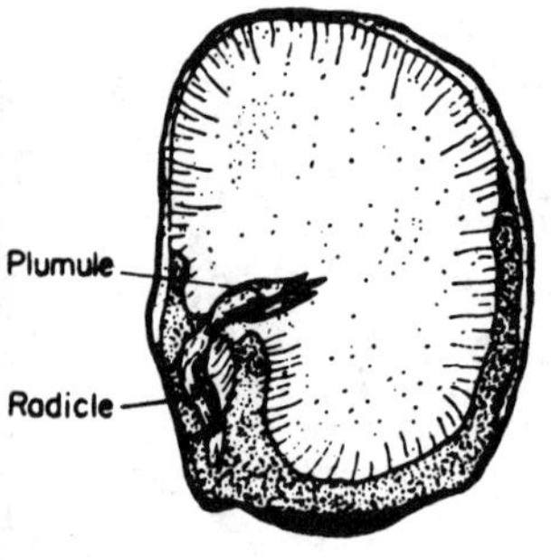

Figure 6 Bean seed with one cotyledon removed.

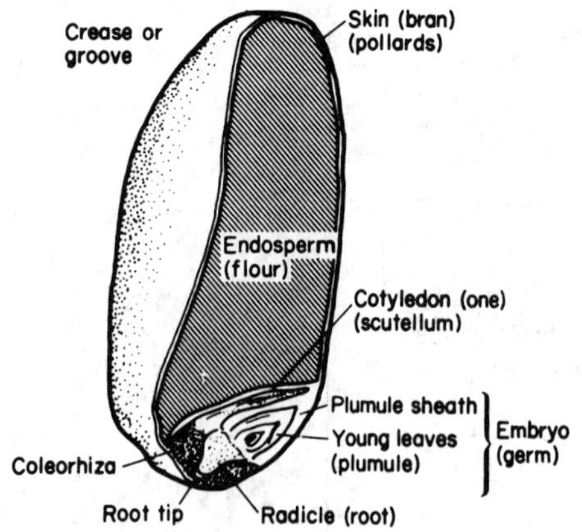

Figure 7 Wheat grain cut in half at the crease.

G. Check your understanding

1. Explain the differences between an annual, a biennial and a perennial.

2. Find the terms in the text which describe the following:
- It attaches a seed to the inside of a pod.
- A small hole at one end of a black scar.
- It separates the embryo from the endosperm.
- It eventually develops into a young shoot.
- A plant that has only one seed leaf.
- It eventually forms the main root of a plant.

3. Explain the following terms:
- Embryo
- Hilum
- Coleorhiza

H. Understanding discourse

1. Listen to this conversation between some students and their tutor. The students should have read the same text that you have just read. The tutor is checking whether they have understood or not. Listen to them talking and say whether the students have answered correctly or incorrectly.

If the answer is correct, put a tick in the box. ☑
If the answer is incorrect, put a cross in the box. ☒

1.	☐	4.	☐
2.	☐	5.	☐
3.	☐	6.	☐

2. Did you notice in that conversation how the tutor asked his questions? If you do not understand something, you can ask for an explanation in a number of ways. Look at this table:

<table>
<tr><td>(Excuse me)
(I'm sorry, but)</td><td>can you
could you</td><td colspan="2">explain . . . .
tell me a bit more about . . . .</td></tr>
<tr><td></td><td colspan="2">I don't really understand . . . .
could you (possibly/please) repeat</td><td>that last bit . . . .
what you just said . . . .</td></tr>
</table>

Imagine you do not understand the following terms. Ask each other for an explanation.
- Coleorhiza
- Plumule
- Micropyle
- Radicle
- Hilum
- Secondary rootlets

UNIT 2

SOILS (1)

A. Understanding a printed text (1)

The following text will introduce you to the topic of **soils**. Look at the way it is divided into sections and paragraphs. Pay attention to the headings and notes in the margins, and to the illustrations and captions.

Now look at these questions:

1. What is this text about?
2. What is the first main section about?
3. What does the diagram illustrate?
4. What are the main types of rock referred to?
5. What two processes are responsible for the breakdown of rocks?

Read the passage through and find the answers to the questions. Remember, you do not have to understand every word to answer the questions.

Soils

How soils are formed

1 Soils are very complex natural formations which make up the surface of the earth. They provide a suitable environment in which plants may obtain *water, nutrients* and *oxygen* for root respiration, and firm *anchorage*. Soils are formed by the weathering of rocks, followed by the growth and decay of plants, animals, and soil micro-organisms. If a farmer is to provide the best possible conditions for crop growth, it is desirable that he should understand what soils are, how they were formed and how they should be managed.

Topsoil

2 The *topsoil* or *surface soil* is a layer about 8–45 cm deep which may be taken as the greatest depth which a farmer would plough or cultivate and in which most of the plant roots are found.

3 Loose, cultivated, topsoil is sometimes called *mould*.

Subsoil

The *subsoil*, which lies underneath, is an intermediate stage in the formation of soil from the rock below.

Soil profiles

4 A *soil profile* is a section taken through the soil down to the parent rock. In some cases this may consist of only a shallow surface soil 10–15 cm on top of a rock such as chalk or limestone. In other well-developed soils (about a metre deep) there are usually three or more definite layers (or horizons) which vary in colour, texture and compaction (see Fig. 25).

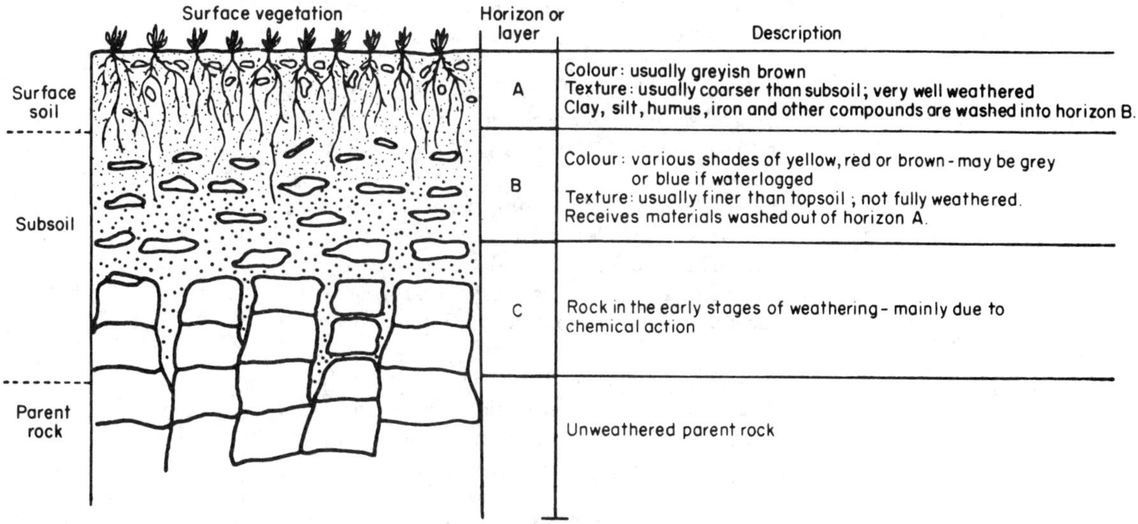

Figure 25 Soil profile diagram showing the breakdown of rock to form various soil layers (horizons).

5 The soil profile can be examined by digging a trench or by taking out cores of soil from various depths with a *soil auger*.

6 A careful examination of the layers (horizons) can be useful in forming an opinion as to how the soil was formed, its natural drainage and how it might be farmed. Some detailed soil classifications are based on soil profile (see Appendix 2).

SOIL FORMATION

Factors which affect soil types

7 There are very many different types of soils and subsoils. The differences are mainly due to the variety of rocks from which they are formed. However, other factors such as *climate, topography, plant and animal life*, the *age of the developing soil material* and *farming operations* also affect the type of soil which develops.

The more important rock formations

Igneous rocks

8 *Igneous or primary rocks*, e.g. *granite* (coarse crystals) and *basalt* (fine crystals). These rocks were formed from the very hot molten material which made up the earth, millions of years ago. The minerals (chemical compounds) in these rocks are mostly in the form of crystals and are the primary source of the minerals found in all our soils. Igneous rocks are very hard and weather very slowly. Clay and sand are breakdown products.

Sedimentary rocks

9 *Sedimentary or transported rocks.* These have been formed from weathered material (e.g. clay, silt and sand) carried and deposited by water and wind. The sediments later became compressed by more material on top and cemented to form new rocks such as *sandstones, clays* and *shales.*

Chalks and limestones

10 The *chalks* and *limestones* were formed from the shells and skeletons of sea animals of various sizes. These rocks are mainly calcium carbonate but in some cases are magnesium carbonate. The calcareous soils are formed from them (see page 29).

Metamorphic rocks

11 *Metamorphic rocks* e.g. *marble* (from limestone) and *slate* (from shale). These are rocks which have been changed in various ways.

Weathering of rocks

The breakdown of rocks is mainly caused by the *physical* and *chemical* effects of the weather.

Physical weathering

13 *Physical weathering. Changes of temperature* cause the various mineral crystals in rocks to expand and contract by different amounts, and so cracking and shattering often occurs.

14 Water can cause pieces of rock surfaces to split off when it freezes and expands in cracks and crevices. Also, the molecules (small particles) of water in the pores and fissures of the rock exert expansion and contraction forces similar to those of freezing and thawing. The pieces of rock broken off are usually sharp-edged, but if they are carried and knocked about by glaciers, rivers or wind, they become more rounded in shape, e.g. sand and stones in a river bed.

15 *Wetting and drying* of some rocks such as clays and shales causes expansion and contraction and results in cracking and flaking.

Chemical weathering

16 *Chemical weathering.* Chemical breakdown of the mineral matter in a developing soil is brought about by the action of water, oxygen, carbon dioxide and nitric acid from the atmosphere; and by carbonic and organic acids from the biological activity in the soil. The soil water, which is a weak acid, dissolves some minerals and allows chemical reactions to take place.

17 Water can also unite with substances in the soil (hydration) to form new substances which are more bulky and so can cause shattering of rock fragments.

This text was taken (edited) from LOCKHART & WISEMAN, *,Introduction to Crop Husbandry,* 5th edn., Pergamon Press

Clay is produced by chemical weathering. In the case of rocks such as granite, when the clay-producing parts are weathered away the more resistant quartz crystals are left as sand or silt.

19 In the later stages of chemical weathering the soil minerals are broken down to release plant nutrients—this is a continuing process in most soils.

Gleying

20 In badly drained soils, which become waterlogged from time to time, various complex reactions (including a reduction process) occur and are referred to as *gleying*. This process, which is very important in the formation of some soils—especially upland soils in the U.K.—results in ferrous iron, manganese and some trace elements moving around more freely and producing colour changes in the soil. *Gley* soils are generally greyish in colour (may also be green or blue) in the waterlogged regions, but rusty-coloured deposits of ferric iron also occur in root and other channels, and along the boundaries between the waterlogged and aerated soil, so producing a mottled appearance (hence the Russian name "gley"). Glazing or coating of the soil structure units with fine clay is also associated with gleying.

B. Check your understanding

Now read the text carefully, looking up any new items in a dictionary or reference book. Then answer the following questions:

1. What are the four main constituent parts of soil?
2. What should the farmer understand about soil?
3. How do you take a soil sample?
4. What can a soil sample tell you?
5. How is sand formed?
6. How is sandstone formed?
7. What are the two main causes of physical weathering?
8. What causes chemical weathering?
9. What is the cause of gleying?
10. What is the normal colour of a gley soil?

C. Increase your vocabulary

In this section you should use your dictionary to help you answer the questions about the text.

1. Look at the first paragraph and say what words have the opposite meaning to:

- simple
- loose
- not wanted
- macro-

2. Look at paragraph 2 and say what these words refer to:

- line 2: which
- line 3: which
- line 4: which

3. Look at paragraphs 3, 4, 5 and 6 again. Which words correspond to the definitions:

- how particles are arranged in a substance
- the process of pressing particles closely together
- a ditch dug in the ground
- the middle, or most important, part of anything
- a system for taking away water
- a group into which something is put

4. Look at paragraph 7 again. Can you explain the words:

- climate
- topography

5. Look at paragraph 8 again.

- What does 'e.g.' mean?
- Is this statement correct or incorrect? Crystals are not found in igneous rocks.

6. Look at paragraphs 9, 10 and 11 again. What words have the same meaning as:

- put down
- pressed together
- of different kinds

7. Look at paragraphs 12, 13 and 14 again. What words have the same meaning as:

- grow larger
- circular
- grow smaller
- happen

8. Look at paragraphs 15 to 19 again. Can you explain the words:

- biological activity
- hydration
- nutrient

9. Look at paragraph 20 again. What words have the same meaning as:

- usually
- edges/limits

D. Check your grammar

EXPRESSING CONDITIONS

Do you remember?
If pests *are* not *checked*, they *will damage* the crop.
Seeds *germinate* if the conditions *are* right.

1. Take one clause from each of the two columns below to make one sentence. Make sure your sentences make sense!

• If you don't understand the soil,	you won't remember the lectures
• If a soil is badly drained,	if the soil is fertile
• Crops are often lost	the world will starve
• If you don't take notes,	if they don't eat proper food
• If agriculture fails,	gleying will occur
• If cows are not fed properly,	it can become waterlogged
• You get better crops	if the farmer does not look after them
• Most plants die	if you don't know how to drive it
• People get sick	you'll never be a good farmer
• A tractor can be dangerous	if they don't get enough water
• If land isn't drained	their milk yield will fall

Do you remember?
If I *had* a glasshouse, I *would grow* ornamental plants.

2. Now say what you think would be the result if the circumstances below actually happened. Complete the sentences, giving your own opinion:

Example:
My car/not/start
If my car didn't start . . . .

- drought/hit/my country
- My land/flood
- I/take/soil sample/near here
- Lose/my notes
- My country's population/double
- There/be/no rice (or wheat)/my country
- There/be/no farm machinery
- We/not test/new plant varieties

Do you remember?
If I *had not chosen* agriculture, I *would have become* a shopkeeper.

3. Use the notes below to make complete sentences on the above model.

- Mankind/not develop/agriculture,
- Animals/not domesticate,
- My tomato crop/fail
- It/not rain/last year
- There/be/a better summer

E. Understanding a lecture

1. You are now going to hear part of a lecture, divided into short sections to help you understand it. As you listen, answer the questions below.

Section 1

- What heading does the lecturer suggest for your notes?
- How many main classes of soil are there?

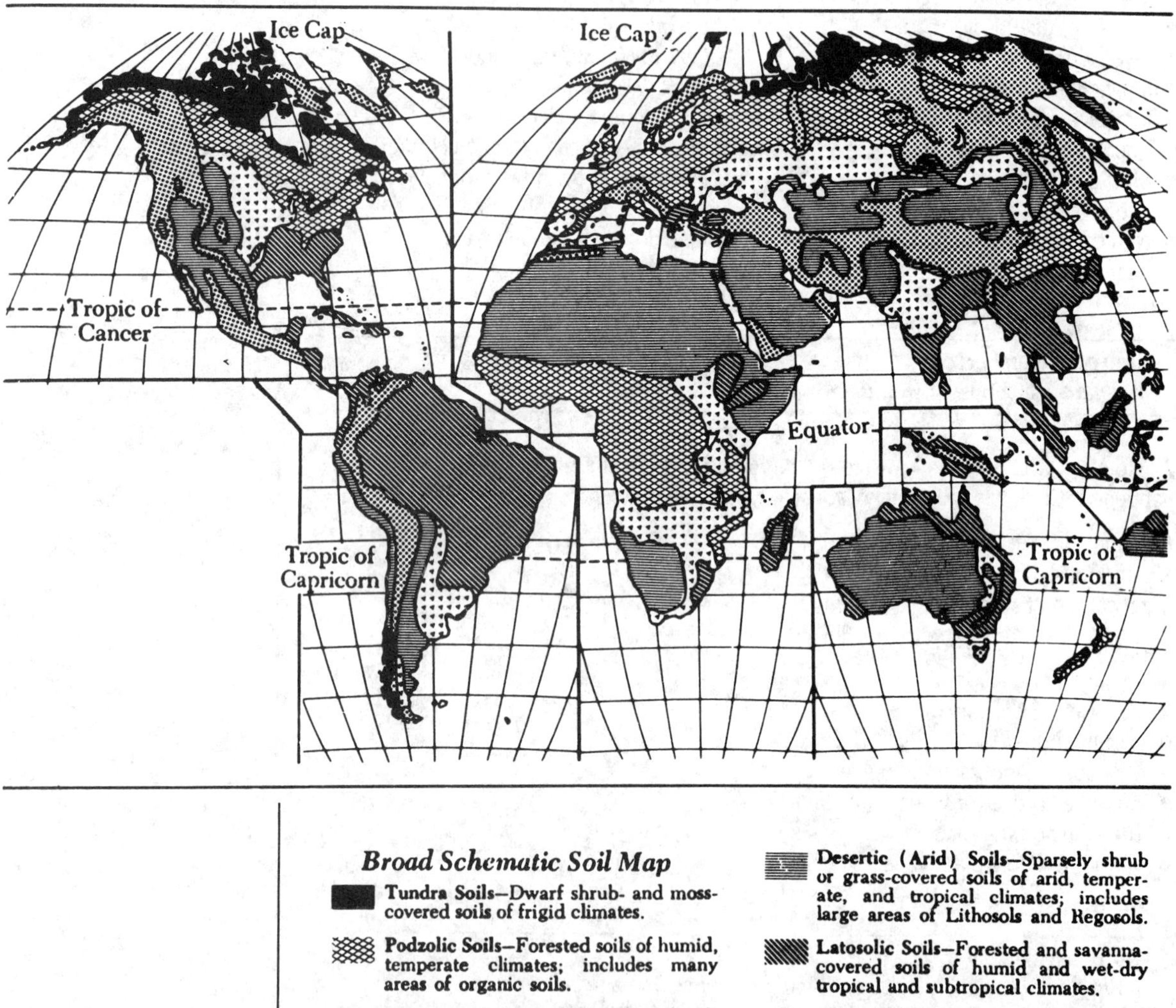

Broad Schematic Soil Map

Tundra Soils—Dwarf shrub- and moss-covered soils of frigid climates.

Podzolic Soils—Forested soils of humid, temperate climates; includes many areas of organic soils.

Chernozemic Soils—Grass-covered soils of subhumid, semiarid temperate climates; includes some soils of wet-dry tropical savannas such as black and dark gray clays.

Desertic (Arid) Soils—Sparsely shrub or grass-covered soils of arid, temperate, and tropical climates; includes large areas of Lithosols and Regosols.

Latosolic Soils—Forested and savanna-covered soils of humid and wet-dry tropical and subtropical climates.

Soils of Mountains—Stony soils (Lithosols) with inclusions of one or more above soils, depending on climate and vegetation, which vary with elevation and latitude.

Section 2

- What is the first kind of soil the lecturer describes?
- You find this kind of soil in climates that are

  very hot ☐
  dry ☐
  cool ☐
  very cold ☐
- Under what kind of tree was this kind of soil formed?
- What colour is this type of soil?
- Is this statement correct or incorrect?

  This type of soil is productive because it is not particularly acid. ☐

Section 3

- What kind of soil is the lecturer talking about?
- What term does the lecturer use to describe soils through which water flows easily?
- Is this statement correct or incorrect?

  This type of soil keeps all its plant nutrients. ☐

Section 4

- Note down one part of the world where you find chernozemic soils.
- Under what type of plant were chernozemic soils formed?
- Is this statement correct or incorrect?

  Chernozemic soils are agriculturally unproductive. ☐
- What colour is this type of soil?

Section 5

- What colour are desert soils?
- Desert soils are this colour because

  They have little organic content ☐
  They have a lot of minerals ☐
  They are not irrigated ☐
  They have never been farmed ☐
- Look at the map. Which types of soil has the lecturer not talked about?

2. Now wind the tape back to the beginning of the lecture and listen to it again. This time, instead of answering questions, take notes. The questions you have already answered will help you do this. When you have listened to the whole of the lecture, you will be asked to provide a short summary of it, using your notes as a guide. So make sure you note down the important points.

F. Understanding a printed text (2)

Read the following text carefully, looking up anything you do not understand.

THE PHYSICAL MAKE-UP OF SOIL AND ITS EFFECT ON PLANT GROWTH

1 The farmer must consider the soil from the point of view of its ability to grow crops. To produce good crops the soil must provide suitable conditions in which plant roots can grow. It must also supply nutrients, water and air; and the temperature must be suitable for the growth of the crop.

2 The soil is composed of:

Solids	Mainly *mineral matter* (stones, sand, silt, clay, etc.) and *organic matter*—remains of plants and animals.
Liquids	Mainly *soil water* (a weak acid).
Gases	*Soil air* (competes with water to occupy the spaces between the particles).
Living organisms	Micro-organisms such as bacteria, fungi, small soil animals, earthworms, etc.

Mineral matter

3 This weathered material, and especially the clay part, is mainly responsible for making a soil difficult or easy to work. It may provide many plant nutrients—but *not* nitrogen. The farmer normally cannot alter the mineral matter in a soil.

4 The amounts of clay, silt and sand which a soil contains can be measured by a *mechanical analysis* of a sample in the laboratory.

Organic matter (o.m.)

5 The organic matter content of a soil can vary from time to time. It usually increases with clay content, e.g. ordinary heavy soils have about 3–4% o.m. compared with 1–1.5% in very light soils: most fertile soils contain about 3–5% (on a dry weight basis). . . . . Organic matter may remain for a short time in the undecayed state and as such can help to "open-up" the soil—this could be harmful on sandy soils. However, the organic matter is soon attacked by all sorts of soil organisms—bacteria, fungi, earthworms, insects, etc. When they have finished eating and digesting it and each other, a complex, dark-coloured, structureless material called *humus* remains: materials produced during the breakdown process are very beneficial in restoring and stabilizing a good soil structure.

6 The amount of humus formed is greatest from plants which have a lot of strengthening (lignified) tissue (e.g. straw). Humus is finally broken down by an oxidation process which is not fully understood.

7 The amount of humus which can remain in a soil is fairly constant for any particular type of soil. The addition of more organic matter often does not alter the humus content appreciably because the rate of breakdown increases. Organic matter is broken down most rapidly in warm, moist soils which are well limed and well aerated. Breakdown is slowest in waterlogged, acid conditions.

8 The chemical make-up of humus is not fully understood but its effects on the soil are well known.

9 Like clay, it is a *colloid* (i.e. it is a gluey substance which behaves like a sponge—it absorbs water and swells up when wetted and shrinks on drying.) The humus colloids are not so gummy and plastic as the clay colloids but they can improve light (sandy) soils by binding groups of particles together. This reduces the size of the pores (spaces between the particles) and increases the water-holding capacity. Humus can also improve clay soils by making them less plastic and by assisting in the formation of a crumb structure—lime must also be present. Earthworms help in this soil improvement by digesting the clay and humus material with lime.

10 Plant nutrients—particularly nitrogen and phosphorus—are released for uptake by other plants when organic matter breaks down. The humus colloids can hold bases such as potassium and ammonia in an available form. In these ways it has a very beneficial effect in promoting steady crop growth.

11 Organic matter in the soil may be maintained or increased by growing *leys*, working-in *straw* and similar *crop residues*, *farmyard manure* (FYM), *composts*, etc. The roots and stubble are usually sufficient to maintain an adequate humus content in a soil growing cereals continuously.

▼

12 In areas where erosion by wind and water is common, mineral soils are less likely to suffer damage if they are well supplied with humus.

13 Where it is possible to grow good leys and utilize them fully, this is one of the best ways of maintaining a high level of organic matter and a good soil structure.

14 Increasing the organic matter (humus) content of a soil is the best way of increasing its water-holding capacity: 50–60 tonnes/ha of well-rotted FYM may increase the amount of water which can be held by 25% or more.

This text was taken (edited) from LOCKHART & WISEMAN, *Introduction to Crop Husbandry*, 5th edn., Pergamon Press

G. Check your understanding

1. Find the terms in the text which mean the following:

- what is left after organic matter has been attacked by soil organisms.
- o.m.
- a substance that expands when wet and contracts when dry
- a space between particles of soil
- the loss of soil due to the action of wind and water
- the remains of corn left in the ground after the stalks have been cut.

2. Which two aspects of humus are not fully understood?

3. Are these statements correct or incorrect?

- The clay part of mineral matter is what makes a soil easy or difficult to work. ☐
- Organic matter cannot harm sandy soils. ☐
- Insects play no part in the formation of humus. ☐
- The amount of air in the soil does not affect the breakdown of humus. ☐
- Humus colloids in a soil mean the soil can hold more water. ☐
- Nitrogen comes from the breakdown of plants. ☐

4. Note down the main benefits of humus to the farmer.

H. Understanding discourse

You are going to your college library for the first time. Your friend Robert is telling you where you can find books and magazines on different aspects of agriculture. As he talks, write down the number or letter of the section where you can find the books you want. You can see the books and magazines you want listed below the diagram.

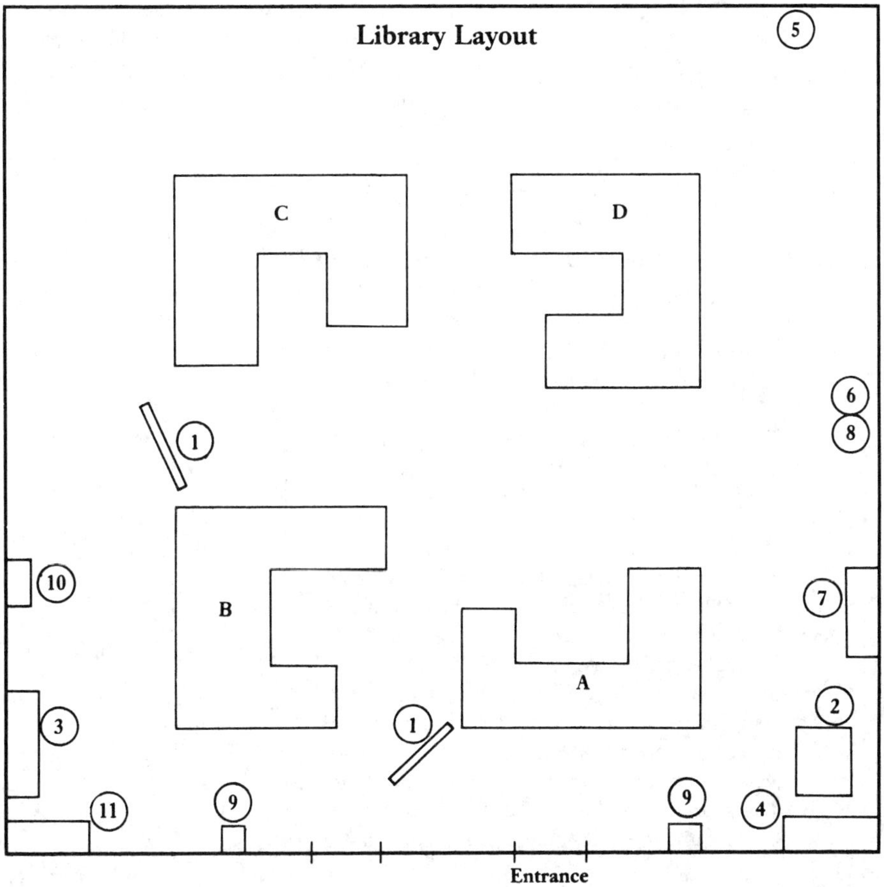

You want to know where to find:

- reference books ____________
- new magazines ____________
- chemistry books ____________
- gardening books ____________
- a book on how to raise hens ____________
- a book on how to raise a dairy herd ____________
- a book on how to irrigate land ____________
- a book on crops ____________
- a book on how to maintain a tractor ____________
- video programmes ____________

UNIT 3 SOILS (2)

A. Understanding a printed text (1)

The following text continues the topic of **soils**. Look at the way it is divided into sections and paragraphs. Pay attention to the headings and notes in the margins, and to the illustrations and captions.

Now look at these questions:

1. What is the text about?
2. What is meant by 'pore space'?
3. Is the pore space the same in all soils?
4. What does Figure 26 illustrate?
5. How is water lost from the soil?

Read the passage through and find the answers to the questions. Remember, you do not have to understand every word to answer the questions.

WATER IN THE SOIL

1 Soils vary in their capacity to hold water, and to understand why, it is necessary to understand some of the differences between soils.

The pore space

2 The soil is a mass of irregular-shaped particles forming a network of spaces or channels called the *pore space*, which may be filled with air or water or both. If the pore space is completely filled with water the soil is *waterlogged* and unsuitable for plant growth because the roots need oxygen for respiration. Ideally, there should be about equal volumes of air and water.

3 When the soil particles are small (e.g. clay) then the spaces between the particles are also small, and when they are large (e.g. sand) the spaces are large. However, although the spaces are small in a clay soil there are very many more spaces than in the same volume of a sandy soil. In a clay soil about half of the total volume is pore space whereas in a sandy soil only about one third is pore space. These volumes refer to dry soils. The pore space may be altered by a change in:

How the pore space can be changed

(a) grouping of the soil particles (i.e. structure),
(b) amount of organic matter (humus) present,
(c) compaction of the soil.

The fact that clay soils have a greater pore space than sandy soils partly explains why the clay soil can hold more water.

Surface area

4 Another important factor is the *surface area* of the particles.

5 Water is held as a thin layer or film around the soil particles. The smaller the particles the stronger are the attractive forces holding the water. Also, the smaller the particles the greater is the surface area per unit volume.

. . . . A comparison for pure materials is set out in Table 3.

TABLE 3

Material	Particle size (mm)	Surface area
Coarse sand	2.0	×
Finest sand	0.02	100 ×
Finest silt	0.002	1000 ×
Finest clay		100,000 ×

6 The surface area of the particles in a cubic metre of clay may be over 1000 hectares.

Organic matter holds water

7 The organic matter (humus) in the soil also holds water.

8 The water in the soil comes from rainfall, or, in dry areas, from irrigation.

Distribution of water in the soil

9 When water falls on a dry soil it does *not* become evenly distributed through the soil. The topmost layer becomes saturated first and as more water is added the depth of the saturated layer increases. In this layer most of the pore space is filled with water. However, a well-drained soil cannot hold all of this water for very long. After a day or so some of the water will soak into the lower layers or run away in drains. The amount of water which is then retained by the soil is called the *moisture-holding capacity* or *field capacity*. The amount of water which can be held in this way varies according to the texture, and structure of the soil (see pages 24 and 25). The weight of water held by a clay soil may be half the weight of the soil particles, whereas a sandy soil may hold less than one-tenth of the weight of the particles. The water-holding capacity of a soil is usually expressed in mm, e.g. a clay soil may have a field capacity of 4 mm/cm in depth.

Ways in which water is retained in the soil

10 The ways in which water is retained in the soil can be summarized as follows:
(a) as a film around the soil particles,
(b) in the organic matter,
(c) filling some of the smaller spaces,
(d) chemically combined with the soil minerals.

The wilting point

Most of this water can be easily taken up by plant roots but as the soil dries out the remaining water is more firmly held and eventually a stage is reached when no more water can be extracted by the plant. This is the *permanent wilting point* because plants remain wilted and soon die. This *permanent wilting* should not be confused with the *temporary wilting* which sometimes occurs on very hot days because the rate of transpiration is greater than the rate of water absorption through the roots; in these cases the plants recover during the night or earlier. The water which can be taken up by the plant roots is called the *available water*. It is the difference between the amounts at field capacity and wilting point.

▼

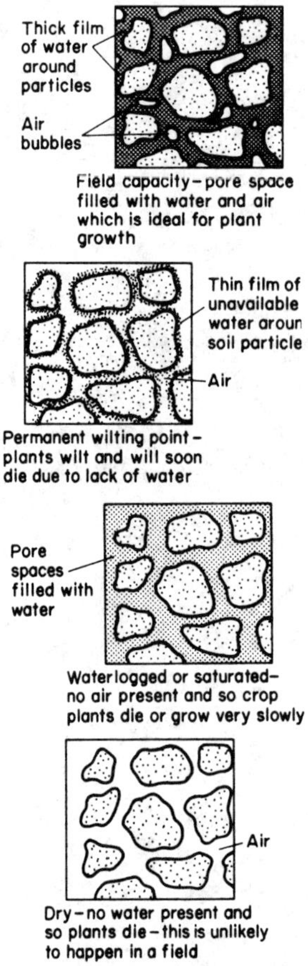

Figure 26 Highly-magnified particles and pores showing how water and air may be found in the soil.

In clay soils only about 50–60% of "field capacity" water is available; in sandy soils up to 90% or more may be available. Although plants may not die until the wilting point is reached, they will suffer from shortage of water as it becomes more difficult to extract (see Fig. 26).

11 Water in the soil tends to hold the particles together and lumps of soil may stick together. When a loam or heavy soil is at or above half field capacity it is possible to form it into a ball which will not fall apart when handled and tossed about. At wilting point, the soil is crumply and will not hold together. So, if irrigation is economically possible, it should be used before the soil dries out to a state in which it will not hold together.

Capillary forces

12 Some of the water in soils with very small pores and channels can move through the soil by *capillary forces,* i.e. surface tension between the water and the walls of the fine tubes or capillaries. This is a very slow movement and may not be fast enough to supply plant roots in a soil which is drying out. Heavy rolling of a soil may reduce the size of the pores and so set up some capillary action.

How water is lost from the soil

13 Water is lost from the soil by *evaporation* from the surface and by *transpiration* through plants. It moves very slowly from the body of the soil to the surface, so after the top 20 to 50 mm have dried out the loss of water by evaporation is very small. Cultivations increase evaporation losses. Most of the available water in a soil is taken up by plants—during the growing season—and air moves in to take its place. This movement of air is easy where the soil has large pore spaces but the movement into the very small pore channels in clay soils is slow until the soil shrinks and cracks—vertically and horizontally—as the water is removed by plants.

14 The water which enters the soil soon becomes a dilute solution of the soluable soil chemicals. It dissolves some of the carbon dioxide in the soil and so becomes a weak acid.

This text was taken (edited) from LOCKHART & WISEMAN, *Introduction to Crop Husbandry*, 5th edn., Pergamon Press

▼

B. Check your understanding

Now read the text carefully, looking up any new items in a dictionary or reference book. Then answer the following questions:

1. Why is waterlogged soil unsuitable for plant growth?
2. Why does clay soil hold more water than sandy soil?
3. How is water held round soil particles?
4. How does water reach the soil?
5. What does the term 'field capacity' mean?
6. How is field capacity of a soil described?
7. What is the 'wilting point'?
8. Why does temporary wilting occur?
9. When should irrigation be used?
10. Why do clay soils shrink and crack?

C. Increase your vocabulary

In this section you should use your dictionary to help you answer the questions about the text.

1. Look at the first two paragraphs again. Which words correspond to the definitions?

- be/become different
- in a way which could not be better
- breathing
- a complex system of lines that cross
- the amount of space occupied by a substance

2. Look at paragraph 3 again. What words have the same meaning as:

- on the other hand
- changed

3. Look at paragraphs 4 to 8 again. Can you explain the words

- surface area
- attractive forces
- hectare
- irrigation

4. Look at paragraph 9 again. What words have the opposite meaning to:

- wet
- gets less
- height
- higher
- lost

5. Look at paragraph 10 again. What words have the same meaning as:

- taken out
- staying
- get better
- in the end

6. Look at paragraph 11 again and say what these words refer to:

- line 8: it
- line 9: it

7. Look at paragraph 12 again.

- What does 'i.e.' mean?

8. Look at paragraphs 13 and 14 again and say what these words refer to:

- line 3: it
- line 9: its
- line 16: It

D. Check your grammar

ADJECTIVES AND ADVERBS

> **Do you remember?**
> The air found in the soil is *really atmospheric* air which has been changed by *various chemical* reactions.

1. Complete the following paragraphs using the words in brackets in their correct adjectival or adverbial form:

Aeration is the replacement of (stagnant) _____ air in the soil with (fresh) _____ air. The process is (main) _____ brought about by the movement of water into and out of the soil. Then, as the surplus water soaks down to the drains or is taken up by plants, (fresh) _____ air is drawn into the soil to refill the pore spaces.

The aeration process is assisted by

(1) changes in temperature
(2) changes in (barometric) _____ pressure.
(3) (good) _____ drainage,
(4) cultivations — (especial) _____ on clay soils,
(5) (open) _____ soil structure.

Sandy soils are (normal, good) _____ _____ aerated because of their (open) _____ structure. Clay soils are (usual, poor) _____ _____ aerated — (particular) _____ when the very (small) _____ pores in such soils become filled with water. (Good) _____ aeration is (vital, important) _____ _____ for germinating seeds and seedling plants.

CHECKING NOTES

Do you remember?
Instead of saying that, he should have said . . .

Now look at the following notes that a student has made on the text. Say if the notes are accurate or not. If they are not, give the correct version.

* A waterlogged soil is one which is ideal for plant growth
* There are two factors which lead to a change in pore space
* There may be more than 1000 hectares of surface area in a cubic metre of clay
* Water falling on a dry soil is evenly distributed
* Moisture-holding capacity and field capacity mean the same thing
* There are three ways in which water is kept in the soil
* Available water means all the water in the soil
* You can form the soil into a ball at wilting point
* Evaporation is one way in which the soil loses water
* Cultivation makes no difference to evaporation losses

E. Understanding a lecture

1. You are now going to hear part of a lecture, divided into short sections to help you understand it. As you listen, answer the questions below.

Section 1

- What is the title of the lecture?
- How many groups is the lecturer going to talk about?

Section 2

- Write down the name of the first type of organism.
- Are these statements correct or incorrect?
  Bacteria have only one cell.
  All bacteria feed on organic matter.
- Bacteria can harm the next crop grown in a field because
  they use nitrogen
  they eat the crop
  they take protein from organic matter
  they provide too much fertiliser

Section 3

- Are these statements correct or incorrect?
  Fungi are not plants
  Fungi feed on organic matter
- What example of a fungus does the lecturer give?
- What conditions do fungi enjoy that bacteria don't?

Section 4

- Write down the name of the type of organism the lecturer talks about.
- Is this statement correct or incorrect?
  There are more of this type of organism than there are bacteria or fungi.
- Note down the type of soil where this type of organism is common?
- What kind of crop can this type of organism damage?

Section 5

- Is this statement correct or incorrect? Algae develop by using carbon dioxide and nitrogen.
- Note down the kind of soil that algae grow well in.
- Why are algae important to rice crops?

Section 6

- What do protozoa and bacteria have in common?
- What do protozoa feed on?

2. Now wind the tape back to the beginning of the lecture and listen to it again. This time, instead of answering questions, take notes. The questions you have already answered will help you do this. When you have listened to the whole of the lecture, you will be asked to provide a short summary of it. So make sure you note down the important points.

F. Understanding a printed text (2)

Read the following text carefully, looking up anything you do not understand.

SOIL TEXTURE AND STRUCTURE

Soil texture

1 *Soil texture* is that characteristic which is determined by the amounts of clay, silt, sand and organic matter which the soil contains. This property normally cannot be altered by the farmer . . . . Soil texture can be measured by a *mechanical analysis* of a sample in the laboratory and classified accordingly (see Table 4) and also by a "feel" test (see p. 260).

Soil structure

2 *Soil structure* is the arrangement of the soil particles individually (e.g. grains of sand), in groups (e.g. crumbs or clods) or as a mixture of the two. It can be altered by: *weather conditions* (e.g. lumps changed to crumbs by frost action or alternate wetting and drying), *penetration of plant roots, cultivations,* etc. It is not possible to measure soil structure satisfactorily, but an experienced person can easily assess its quality at any one time by its appearance and "feel".

3 It is possible to classify the texture of a soil by rubbing a moist sample of it between the thumb and fingers:

(1) Clay is sticky, will take a polish, and can be moulded.
(2) Silt feels silky, smooth and slightly sticky.
(3) Sand feels gritty.
(4) Organic matter usually feels soft and slightly sticky.

The dominant "feel" indicates the texture group of the soil; if there is no dominant feeling then the soil is a loam.

Heavy and light soils

4 Soils are often classified by farmers and others as heavy, medium and light (not a weight measure). The terms "heavy" and "light" refer to the amount of power required to draw a plough or cultivator through the soil. A heavy (clayey) soil consists mainly of very small particles which pack tightly together whereas a light (sandy) soil consists mainly of large particles which are loosely held together because of the relatively large pore spaces.

Crumb structure

5 Crumb structure is formed by the grouping together (aggregation) of the particles of clay, sand and silt. This aggregation is possible because there are positive and negative electric charges (forces) acting through the surface of the particles. These forces are strongest in clay and very weak in sand. This strong adhesive property of clay particles makes clay soils more difficult to work than sandy soils but it also enables them to form crumbs easily.

6 Water has special electric properties and its presence is necessary for the grouping (crumbing) of soil particles. The electric forces in the water and in the soil particles make the water stick as a thin film around the particles of soil. As this film becomes thinner (e.g. when soil is drying out) the particles are drawn closer together to form groups (crumbs). The particles in the crumbs may come apart again if the soil becomes very wet.

7 There must be lime present in the water if clay particles are to stick together to form porous crumbs. This partly explains why liming benefits clay soils.

8 If organic matter or an iron compound (ferric hydroxide) is present then the particles in the crumbs may remain cemented together and have a more lasting effect on soil structure. Too much ferric hydroxide can have a harmful effect because tightly cemented crumbs are very difficult to wet again after they have dried out.

9 Where there is very little organic matter or ferric hydroxide the stability of the crumbs depends mainly on the amount of clay present. The more clay there is, the stronger will be the forces holding the particles together.

Stability of soil structures

10 Some soil structures are more stable than others, e.g. clays usually have a more stable structure than silts. Soils containing fine sand and silt easily lose their structure and are difficult to work if they are low in organic matter. This is because under wet conditions the sand and silty materials flow very easily and block the aeration and drainage channels in the soil.

▼

Tilth

11 *Tilth* is a term used to describe the condition of the soil in a seed bed. For example, the soil may be in a finely-divided state or it may be rough and lumpy; also, the soil may be damp or it may be very dry. Whether a tilth is suitable or not partly depends on the crop to be grown. In general, small seeds require a finer tilth than large seeds.

SOIL FERTILITY AND PRODUCTIVITY

12 Soil fertility is a rather loose term used to indicate the potential capacity of a soil to grow a crop (or sequence of crops). The productivity of a soil is the combined result of fertility and management.

Natural fertility

13 The fertility of a soil at any one time is partly due to its natural make-up (inherent or *natural fertility*) and partly due to its *condition* (variable fertility) at that time.

14 *Natural fertility* has an important influence on the *rental* and *sale value* of land. It is the result of factors which are normally beyond the control of the farmer, such as:

(1) the texture and chemical composition of the mineral matter,
(2) the topography (natural slope of the land)—this can affect drainage, temperature and workability of the soil,
(3) climate and local weather—particularly the effects on temperature, and rainfall (quantity and distribution).

Soil condition

15 *Soil condition* is largely dependent on the management of the soil in recent times. It can be built up by good husbandry but if this high standard is not maintained the soil will soon return to its natural fertility level. The application of fertilizers can raise soil fertility by increasing the quantities of plant food in the growth and decay cycle.

Management

16 *Management* can control the following production factors:

(1) the amount of organic matter in the soil (see *"Soils"* pp. 20, 21),
(2) artificial drainage and irrigation (see *"Soil Improvement"* pp. 33–44),
(3) erosion (removal of soil by wind and water) (see *"Claying"* pp. 44 and 49),
(4) pH of the soil (see *"Liming"*, p. 31), and the plant nutrients applied (see *"Fertilizers"* p. 53 and *"Crops"* Chapter 4),

This text was taken (edited) from LOCKHART & WISEMAN, *Introduction to Crop Husbandry*, 5th edn., Pergamon Press

▼

(5) cultivations and time of planting (see *"Cultivations"*, pp. 44–51),
(6) variety and plant spacing (see individual crops),
(7) sequence of cropping (see *"Rotations"*, pp. 71–73),
(8) weeds, pests and diseases (see separate chapters),

17 Good management of the above factors should maintain or increase soil fertility and at the same time be commercially profitable. These subjects are dealt with in more detail in other chapters.

G. Check your understanding

1. Explain in your own words the terms:
- soil texture
- soil structure

2. What factors can change the structure of a soil?

3. Can you explain what is meant by the 'feel' of a soil?

4. Can you explain why clay soils are harder to work than sandy soils?

5. What is 'tilth'?

6. What two factors decide the productivity of a soil?

7. Can you explain the term 'natural fertility'?

8. Explain in your own words why management of the soil can affect soil condition.

H. Understanding discourse

A lecturer has been talking about the approximate mechanical analyses of different types of soil. Below you can see the table that a student drew up while he listened to the lecturer. The student has got some of the information wrong. Listen to the lecturer, and correct the student's table.

TEXTURAL GROUPING OF SOILS
(ON WET WEIGHT)

Soil type	*Texture*	*Clay (%)*	*Silt (%)*	*Sand (%)*
Clay	Fine (heavy)	up to 50	15–25	over 35
Clay loam	Fine (light)	30–50	15	35–45
Loam	Medium	20–30	50	30–35
Sandy loam	Coarse (heavy)	10–20	15–25	55–75
Sand	Coarse (light)	0	0	80–100

UNIT 4

CEREALS

A. Understanding a printed text (1)

The following text will introduce you to the topic of **cereals**. Look at the way it is divided into sections and paragraphs. Pay attention to the headings and notes in the margins, and to the illustration and captions.

Now look at these questions:

1. What kind of food crop is this text about?
2. Apart from cereals, one food plant is named as important worldwide. Which is it?
3. Did any food plants originate in both the new and old worlds?
4. How is rice usually classified?
5. What does the table in Figure 20.3 tell you?

Read the passage through and find the answers to the questions. Remember, you do not have to understand every word to answer the questions.

The 'big three'

FOOD CROPS

1 The greater proportion of man's food is derived from relatively few plant species. Most of the staple crops are members of the grass family that yield grains. Recent statistics show rice, wheat, and corn closely grouped as the "big three" in world cereal production. Each accounts for essentially one-quarter of the world's total cereal supply of more than a billion metric tons annually. Barley, oats, rye, millets, and sorghum together make up the remaining quarter.

2 Only one other food plant, the potato, a root crop, gives the cereals much of a run so far as world importance is concerned. Total tonnage of the potato is approximately equal to that of the leading grains but since the potato is fleshy and consists of only 22 percent dry matter, its total food value is considerably less. Sweet-potatoes and yams, various seed legumes (beans, peas, soybean, peanut, etc.), and sugar cane are the other major noncereal food crops. For all practical purposes man relies upon this handful of crop plants for his existence.

People rely on a small number of food plants

3 The modern world would grind to a standstill if it could not depend upon the gradual improvement of varieties of these domesticates, which lend themselves so well to modern mechanized agriculture. Many thousands of wild plants were utilized by primitive cultures; some of them, of course, were used only in times of dire need. Table 20-1 categorizes a number of important crops according to whether they originated in the Old World or the New World. Since the fifteenth century these species have been spread widely over the globe, wherever they proved adaptable to local climate. Area of origin and major zone of today's production do not often coincide!

Table 20-1 NEW WORLD AND OLD WORLD FOOD PLANTS BEFORE 1492.

NEW WORLD		OLD WORLD		
maize	pumpkin	wheat	onion	wine grape
manioc	peanut	rye	garlic	apricot
lima bean	chayote	barley	spinach	peach
potato	papaya	oat	eggplant	olive
sweet-potato	avocado	millet	lettuce	fig
kidney and other beans	pineapple	sorghum	endive	almond
tomato	custard apple	rice	celery	quince
green pepper	soursop	buckwheat	asparagus	pomegranate
Jerusalem artichoke	cherimoya	turnip	peas, various	watermelon
sunflower	guava	cabbage	soybean	cucumber
quinoa	chocolate	rutabaga	broad bean	banana
squash	cashew	chard	dasheen	orange
	sapote	mustard	yam	lemon
		radish	apple	lime
		beet	pear	date
		parsnip	plum	mango
		carrot	cherry	clover
		breadfruit	alfalfa	
		mangosteen	bluegrass	

THE CEREALS

Cereals provide half the world's food needs

4 Cereals are the staff of life to civilization. Around 70 percent of the world's harvested acreage of about a billion hectares, is devoted to growing them. They are the direct source of half the food needs of the world, and the secondary source of much additional food when converted to meat, milk, eggs, and other animal products.

The major cereals

5 The true cereals are all members of the grass family, the Gramineae. The fruit they yield is a grain, a type of fruit in which the ovary wall turns hard and durable, fusing with the single seed (Fig. 20-1). Buckwheat, amaranth, and a few similar seeds of other families ("pseudo-cereals") are sometimes regarded as grains because their seeds are quite similar to those of grasses. This section will be devoted to the true cereals, of which the major crops are rice (*Oryza*), wheat (*Triticum*), maize (*Zea*), barley (*Hordeum*), oats (*Avena*), rye (*Secale*), sorghum (*Sorghum*), various millets (*Panicum, Setaria*, etc.), and a few other species of lesser world importance.

Why cereals are so important

6 There are a number of reasons why cereals have evolved into man's leading food source. Most of them are annuals, or are at least adapted to cultivation as annuals, permitting facility in cropping, whether on the midden of an early domestication center or as a component of today's highly mechanized farming. Grasses in general are quite versatile, too; they adapt well to a variety of soils, climates, and ways of handling. They are also relatively efficient in garnering the sun's energy, transforming it into usable food substance. They are generally hardy and recuperative, and are plagued by no more than their share of diseases and pests. But above all, the grain is a neat package of stored energy, conveniently harvested, easily cleaned and handled, amenable to storage without special drying.

▼

7 Rice, wheat, and corn are the world's three major cereals, all about equally important in terms of world production. Rice is the prime source of sustenance for tropical populations; it is grown mostly on flood plains or where the land can be seasonally ponded. Wheat is chiefly grown on lands that were naturally prairies.

RICE, ORYZA SATIVA

Where rice is grown

8 Rice, or paddy, widely grown in Asia from Afghanistan to the East Indies and north to Japan is the principal food for 60 percent of mankind. It is also a staple in parts of Africa, South America, and to some extent in the southern United States. World production exceeds 300 million metric tons annually, mostly from the Far East (China, India-Pakistan, Japan, Indonesia, and Thailand are the major producing countries).

Classification of rice

9 For practical purposes rices are generally classified as long-grain, medium-grain, or short-grain. The long-grain types need a longer season to mature, and hence are grown mostly in tropical regions, such as Southeast Asia. Connoisseurs prefer the long-grain type, which is nonglutenous and does not turn so soft and sticky when cooked. Medium-grain rices, somewhat softer, are most frequently grown in America. Short-grain types, even more starchy when cooked, are well adapted to more northerly climates, such as that of Japan.

Methods of growing rice

10 Most rice is grown in paddy fields of standing water. There are, however, various forms of upland rice, cultivated in much the same way as any small grain. Upland rice yields tend to be lower, and its importance is not great; for example, in Indonesia, more than 90 percent of the rice is grown in irrigated paddy land. The best rice land is level, conveniently diked for alternating irrigation and drainage.

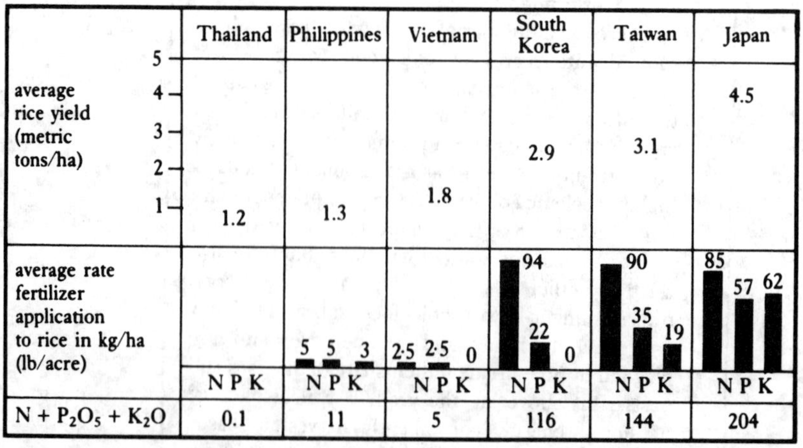

Figure 20-3 Rice yields related to fertilizer usage in several Asiatic production areas. [After Kemmler, World Crops **14**:*117, 1962.*]

11 Japan practises the most progressive rice growing in the Orient. Yields average nearly 4 tons per hectare and can be as high as 10 tons. Fertilization in Japan is as intensive as on North American corn land (Fig. 20-3). A complete fertilizer is generally "plowed down" prior to flooding and hand planting of the seedling rice plants. Another "top-dressing" is applied just before heading to increase yields. The fields are drained for harvesting and threshing, typically accomplished with small portable machines on the site. A green manure (vetch) is sometimes planted and plowed down. The paddy fields may also be used for a winter crop, such as barley, wheat, or rape.

Advanced methods of growing rice

12 Under more primitive circumstances, such as those that prevail in many of the tropical lands, rice growing is an exceedingly laborious and time-consuming hand operation. As noted in Figure 20-3, even fertilization is quite limited in many places, although compost, and sewage have been used since ancient times to fertilize the paddy. After the paddy is flooded to soften the soil, the ground is worked, often by water buffalo pulling a forked stick (small hand tractors may be used in more advanced countries, such as Japan). The rice seedlings are started elsewhere, sometimes under government auspices, and are transplanted by hand into flooded paddies, usually between November and January in southeast Asia, to take advantage of the winter monsoon. About four months later the paddies are drained and the stand harvested – another laborious job, accomplished chiefly by women. The rice stalks are cut individually with a small sharp sickle, banded together, and laid out in the sun to dry. If there is sufficient water, either a second rice crop or some other crop may then be planted to the same land.

Less advanced methods

13 Various means are employed to clean and dehull newly harvested rice. In some countries the sun-dried heads are threshed in the field by treading, either by man or live-stock (of course, with considerable loss of grain). In Afghanistan green rice is mixed with heated sand to harden and "crisp" the hulls, which are then separated by crude milling on water-powered devices that pound the grain against the soil. Screening and winnowing follow. In Ecuador much of the rice is fermented by massing it on the floor while damp and covering it with a tarpaulin for a few days. Upon subsequently drying it cooks more quickly than unfermented rice in the high altitudes of the Andes. Fermentation may also provide healthful by-products.

Cleaning rice

This text was taken from JANICK *et al.*, *Plant Science*, Freeman, San Francisco

B. Check your understanding

Now read the text carefully, looking up any new items in a dictionary or reference book. Then answer the following questions:

1. Which are the 'big three' cereals?
2. Why is the potato less valuable than cereals?
3. Which of the food plants given in Table 20.1 is important in your country?
4. What are the Gramineae?
5. Why is buckwheat called a 'pseudo-cereal'?
6. Which is the most important reason, in your opinion, for cereals being our leading food plant?
7. Which is more important to the world, rice, wheat or corn?
8. Why do some people prefer long-grained rice?
9. Summarise the less advanced method of growing rice as described in the text.
10. Why is fermented rice more suitable for people living in the Andes?

C. Increase your vocabulary

In this section you should use your dictionary to help you answer the questions about the text.

1. Look at the first paragraph again. What words have the same meaning as:

- yearly
- obtained from
- give/provide

2. Look at paragraph 2 again. What words in the text could you replace with:

- depends on
- about
- small number
- contains

3. Look at paragraph 3 again. What words have the opposite meaning to:

- quick
- modern
- hand-operated
- domesticated

4. Look at paragraph 4 again. What words have the same meaning as:

- extra
- changed into

5. Look at paragraph 5 again. Which words correspond to the definitions?

- something that lasts a long time
- becoming joined
- the seed vessel in a plant

6. Look at paragraphs 6 and 7 again. Make the opposite of these adjectives, using the prefixes *un-* and *in-* whenever possible.

- convenient
- hardy
- important
- early
- usable
- mechanised
- easy
- efficient
- natural
- high
- major
- equal

7. Look at paragraphs 8 and 9 again. Can you explain the words:

- staple
- metric ton
- tropical
- season
- starchy

8. Look at paragraphs 10 and 11 again. Which words correspond to the definitions?

- can be carried
- beating the grain out of a crop
- animal waste used as a fertiliser
- taking place by turns
- the higher parts of a country

9. Look at paragraphs 12 and 13 again. Can you explain the words:

- sickle
- compost
- seedling
- transplant
- hull
- milling
- screening
- winnowing

D. Check your grammar

WRITING SENTENCES

> **Do you remember?**
> Some verbs in English are followed by the infinitive, 'to':
> I *want to finish* my essay today.
>
> Others are followed by the -ing form:
> I *enjoy playing* football.
>
> Verbs followed by a particle like 'in', 'from', etc. are followed by the -ing form:
> He *insisted on going* to a film last night.
>
> Some verbs can take both 'to' and -ing, but their meaning sometimes changes:
> He *remembered* (= did not forget) *to bring* his books.
> I *remember* (= recall) *meeting* his brother last year.

1. Complete the following sentences:

- He suggested (watch) .... the football match on Saturday.
- I didn't stop (work) .... until midnight last night.
- I'm sorry, I've forgotten (bring) .... my essay.
- What machine do we use (plough) .... the fields?
- You must avoid (spill) .... chemicals on your hands.
- Try (understand) .... what the lecturer is saying.
- Try (look) .... that word up in your dictionary.
- Would you mind (explain) .... that again?
- The lecturer began (talk) .... about cereal crops.
- Tomorrow I intend (discuss) .... different kinds of fertiliser.
- Sickness prevented him from (attend) .... the demonstration.
- I am looking forward to (study) .... agriculture.

WRITING PARAGRAPHS

2. Match the following clauses together, using *and, but, when, after, so* and *because.*

• the next evening he needed them	Robert could not attend
• he arrived at the canteen	Robert finally arrived
• this did not please Andrew very much	he went to the canteen to look for Robert
• last night Andrew lent Robert his lecture notes	Robert had not returned them
• Andrew needed his notes immediately	Robert was not there
• he had waited for an hour	he decided to wait

3. Now arrange your sentences into a sensible paragraph.

E. Understanding a lecture

1. You are now going to hear part of a lecture, divided into short sections to help you understand it. As you listen, answer the questions below.

Section 1

- What crop did the students study last week?
- Which crop are they going to study this week?
- Write down the Latin name for this crop.

Section 2

- If you farm on a prairie, the most suitable crop is *(tick the correct answer):*
  maize ☐
  wheat ☐
  rice ☐
- Note down three characteristics of prairies that the lecturer describes.
- Note down the percentages that the lecturer gives you on the diagram below.

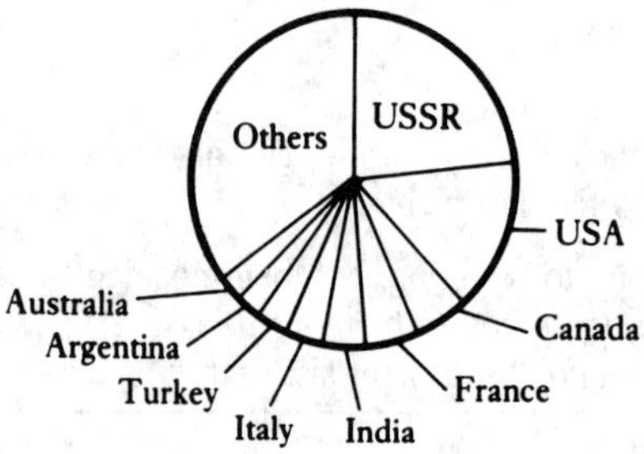

Section 3

- Are these statements correct or incorrect?
  Wheat was used in Iraq approximately 9,000 years ago. ☐
  Wheat was also used in Egypt 9,000 years ago. ☐
  Bread was eaten in Egypt about 4,500 years ago. ☐

Section 4

- Note down the two classes of wheat the lecturer mentions.
- Which class of wheat has more protein?
- Which class of bread is used for pastries?

Section 5

- Winter wheat is sown in *(tick the correct answer):*
  spring ☐
  summer ☐
  autumn ☐
  winter ☐
- What period of time is the lecturer referring to when he mentions '90 days'?

2. Now wind the tape back to the beginning of the lecture and listen to it again. This time, instead of answering questions, take notes. When you have listened to the whole of the lecture, you will be asked to provide a short summary of it. Make sure you note down the important points.

F. Understanding a printed text (2)

Read the following text carefully, looking up anything you do not understand.

Where maize is produced

Maize, Zea mays

1 Maize (Fig. 20-7), . . . a valuable find from New World exploration . . . . now makes its mark world-wide. But in no part of the world is maize cultivated so intensively as in the United States, where half of the annual quarter-billion tons of corn grain is grown. Nor have many crops attained the degree of sophistication that has come to hybrid corn production in North America. A distant second to the United States in maize production is usually Brazil, with the U.S.S.R., Mexico, Romania, Yugoslavia, Argentina, India, and South Africa following more or less in that order.

Fig. 20-7

The origins of maize

The spread of maize

Modern cultivars

2 Maize appears to have been domesticated in the Tehuacan Valley of Mexico. Maize pollen 80,000 years old has been collected from cores taken from old lake beds near Mexico City, which certainly proves it indigenous to the region. The original wild form has probably long been extinct. Evidence suggests that cultivated maize arose through natural crossing, perhaps first with gama-grass, *Tripsacum dactyloides* (which itself may have originated from a cross of primitive *Zea* with *Manisurus*), to yield Teosinte, *Euchlaena mexicana*, and then possibly with backcrossing of Teosinte to primitive maize to produce modern races. The place of origin has long been debated; if it truly were southern Mexico, which is an excellent possibility, maize soon spread throughout the Americas to become a mainstay food. It might easily have been carried southward into the foot-hills of the Andes, east along the northern coast of South America, and thence northward from one Caribbean island to another and on to eastern North America. At the same time it most certainly spread overland northward from Mexico into the western United States, and adapted forms spread northeastward as far as New England.

3 Modern cultivars of maize resemble the primordial form only remotely. Indeed, modern maize is wholly a ward of man, unable to survive and perpetuate itself without his care in harvesting and planting. Man's selection through the millenia has yielded a specialized and very queer looking crop, with the female (pistillate) flowers concentrated on one or a few lower cobs, or ears, and the male (staminate) flowers on separate, terminal tassels. The seeds (grains on the cob) shatter little and because they are enclosed in husks they have scant chance for natural dispersal. Growing of maize today depends upon servicing this odd creation with a variety of contrivances to handle its soil preparation (cultivation, fertilization), planting (planter drills), care (pest control), harvesting (picker-shellers, combines), and cleaning and storing (drying equipment, grain elevators). Add to this all the special accoutrement for hybrid seed production – inbreds, male-sterile lines, detasseling equipment, and so on. Here is a remarkable marriage of biology with technology – interlocking responses and mutual dependence between plant and man.

This text was taken from JANICK *et al.*, *Plant Science*, Freeman, San Francisco

▼

Early methods of planting

4 Early planting of maize by the American Indians was, of course, much more casual. But even by then selections had been made, and this subtropical species had been adapted to a wide variety of habitats, yielding such modern types as popcorn.

5 . . . . The Indians planted corn either in slash-and-burn forest clearings, or on drier lands that were seasonally moist or irrigated. Planting was accomplished by such simple techniques as poking holes into the soil and inserting seeds. At least with the Mohawk Iroquois, the women soaked the corn seed for a few days before planting a few to a hill. In certain locations primitive fertilization was practiced, such as burying fish parts at planting. Other crops, wuch as beans and squash, were often interplanted with the maize.

G. Check your understanding

1. Find the terms in the text which describe the following?

- unable to reproduce
- a plant derived from different species or varieties
- producing a plant by mixing species or varieties
- a machine that both reaps and threshes

2. Can you explain the following words:

- plant drill
- pistillate
- staminate
- slash-and-burn

3. These words can be explained in simpler, more everyday language. Can you do that?

- contrivance
- accoutrement
- accomplish
- mainstay
- terminal

4. Now look at these statements. Using the information in the text, say if they are correct or incorrect.

- Brazil produces nearly as much maize as the USA.
- Maize was imported into Mexico many thousands of years ago.
- Cultivated maize probably developed from natural crossing.
- Maize has never been cultivated on the Atlantic coast of America.
- Modern types of maize are very similar to the original types.
- Maize cannot survive as a plant without man's help.

H. Understanding discourse

Listen to this lecturer explaining what you must do if there is a fire in the building. You are in room 401. If there was a fire now, where would you go?

UNIT 5

LIVESTOCK (1)

A. Understanding a printed text (1)

The following text will introduce you to the topic of **livestock**. Look at the way it is divided into sections and paragraphs. Pay attention to the headings and notes in the margins, and to the illustrations and captions.

Now look at these questions:

1. What is the purpose of this text?
2. Why are carbohydrates important?
3. What is the importance of proteins?
4. What does Figure 1 show you?
5. What does Table 1 show you?

Read the passage through and find the answers to the questions. Remember, you do not have to understand every word to answer the questions.

FEEDING FARM ANIMALS

The constituents of foods

Composition of feedingstuffs

1 Food accounts for the major portion of the production costs of all animal products, and errors in feeding can seriously reduce the farmer's profits. A sound knowledge of foods, and of rationing, therefore, forms an integral part of the art of stockmanship.

2 A study of feedingstuffs reveals that they are made up as follows:

Food < Water / Dry Matter (Carbohydrate, Fats and Oils, Proteins, Minerals, Vitamins)

3 The feeding value of the dry matter varies considerably between different foods, and in certain cases it can be of more value to some classes of stock than to others.

The importance of carbohydrates

4 *Carbohydrates* supply most of the energy which farm animals require to enable their muscles to do work, to keep themselves warm and, for example, to produce milk. They are compounds of carbon, hydrogen and oxygen.

5 The simplest carbohydrates are the sugars. An example is *glucose* which plants can produce in their leaves. To do this they combine water from the soil, carbon dioxide from the air and the sun's energy. Plants then build this *glucose* up into more complex carbohydrates such as *starch*, for storage, or *cellulose*, which forms the young plant's cell walls.

6 As plants become older their cell walls may be strengthened by *lignin*. This largely accounts for the very fibrous nature of old plants. The stems of old kale and of hay made from very mature grass are typical examples of lignified foods.

7 The *cellulose* and *lignin* within a food are frequently referred to together as the *fibre*. In some plants, for instance in old kale and even celery, part of this *fibre* can readily be distinguished by its stringy nature.

8 In general, when plants become old, and their fibre content increases, they reduce in feeding value. This is because a large proportion of the fibre is indigestible.

9 Cows and sheep have a digestive system which is adapted to deal with fibre.

10 Most animal feedingstuffs are derived from plants and, consequently, our livestock are almost entirely dependent upon plants for their supply of energy.

The dependence of animals on plants

Fats and oils

11 *Fats and oils* are chemically very similar to each other, but at normal temperatures fats exist in a solid state and oils are liquids. Like carbohydrates, they contain carbon, hydrogen and oxygen, but they differ in the proportions in which these elements are combined.

Glycerol and fatty acids

12 Just as the simplest "building blocks" in the carbohydrates were the sugars, such as *glucose*, the simplest "building blocks" in the fats are *glycerol* and *fatty acids*.

13 Weight for weight, fats contain more than twice the energy of carbohydrates. When animals have a surplus of energy they convert most of it into fat for storage. This can then be re-utilised if required in times of need.

14 Fats and oils in the diet are also used as a source of energy, and can be converted into animal fat. Some feedingstuffs, for example flaked maize, contain oils which cause animals to lay down soft fat.

The effect of too much fat

15 Excessive fat in the diet may lead to digestive troubles. Consequently, although fat is very rich in energy, most of the energy requirements of farm animals are supplied by carbohydrates.

How animals use protein

Proteins

16 *Proteins* are required by animals to build up lean meat or muscle. They also use them to form internal organs, hair, wool, skin and to repair damaged body tissues. In addition, the milking animal will require protein to help produce milk.

17 Like carbohydrates and fats, they contain carbon, hydrogen and oxygen. However, they also contain nitrogen and some contain sulphur, phosphorus or iron.

Amino acids

18 There are many different types of protein, but they are all built up from different combinations of simpler substances called *amino acids.* Over twenty *amino acids* are known to be important to farm animals. Stock produce many of them in their own bodies by conversion from other *amino acids* or from nitrogen containing substances within their food (Fig. 1)... ... ... These are called the *essential amino acids*. Foods which contain a wide range of these essential amino acids are said to be of *high biological value.*

The danger of too much protein

19 Care should be taken not to overfeed stock with protein. They have the ability to extract the nitrogen and liberate the energy, but this puts a strain on the liver and kidneys. Protein-rich foods are also normally more expensive than those rich in carbohydrates. This is principally because a large proportion of our protein foods, such as groundnut cake, have to be imported.

▼

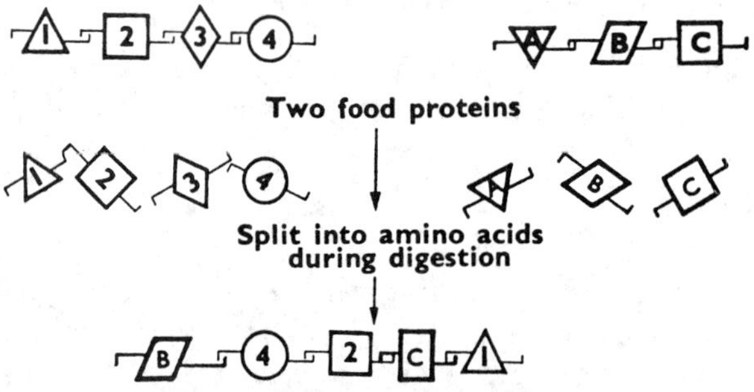

Minerals

20 *Minerals* also form an important part of an animal's diet since they play a vital role in a very large number of the body's normal functions. Those minerals which are required in comparatively large quantities are known as *essential minerals* (Table 1).

TABLE 1

The Functions Associated with the Essential Minerals

Calcium, phosphorus	Bone and teeth formation Growth of soft tissues Fertility
Magnesium	Function of the nervous system Bone formation
Iron	Blood formation
Sodium, potassium	Regulation between acidity and alkalinity of body fluids Food digestion
Sulphur	Part of certain proteins, e.g. hair protein

The importance of minerals to young animals

21 Although minerals only constitute about 3% of the total body, about 80% of the mineral matter, chiefly calcium and phosphorus, is found in the skeleton. It is, therefore, particularly important to ensure that young animals whose skeleton is growing rapidly obtain a good supply of minerals.

22 The types and quantities of minerals found in different feedingstuffs are extremely variable. Animals feeding on certain diets may, therefore, need supplementary mineral rations to prevent deficiency diseases. A wide range of mineral supplements is available from commercial firms, although home mixes using a proportion of steamed bone flour, ground chalk and common salt are sometimes used to supply the essential minerals.

23 Other minerals which are only necessary in small quantities are called *trace elements*. These include iodine, manganese, cobalt, copper and zinc.

This text was taken from BUCKETT, *Introduction to Livestock Husbandry*, 2nd edn., Pergamon Press

▼

B. Check your understanding

Now read the text carefully, looking up any new items in a dictionary or reference book. Then answer the following questions:

1. Does all dry matter have the same feeding value?
2. What are carbohydrates made up of?
3. Why are some older plants fibrous?
4. Why do fibrous plants have less feeding value?
5. Which contain more energy, fats or carbohydrates?
6. Why can too much fat be undesirable?
7. How are proteins built up?
8. Which foods are said to be of 'high biological value'?
9. Which are the essential minerals?
10. What are trace elements?

C. Increase your vocabulary

In this section you should use your dictionary to help you answer the questions about the text.

1. Look at paragraphs 1 to 4 again. What words in the text could you replace with:

- mistakes
- cut down
- shows
- a great deal
- provide

2. Look at paragraphs 5 and 6 again and say what these words refer to:

- line 2: this
- line 2: they
- line 7: this

3. Look at paragraphs 7 to 10 again. What words have the same meaning as:

- as a result
- often
- for example

4. Look at paragraphs 11 to 15 again. Which words correspond to the definitions:

- joined together
- the sort of food usually eaten by an animal
- too much
- changed from one form to another

5. Look at paragraphs 16 to 19 again. What words have the opposite meaning to:

- put in
- fat
- external
- unusually
- cheap

6. Look at paragraphs 20 to 23 again. Can you explain the words:

- skeleton
- deficiency disease
- commercial firms

D. Check your grammar

Do you remember?
That demonstration *hasn't taken place* yet. It's tomorrow.
The lecture *has* already *been given*. It was yesterday.

1. Look at this timetable for an Agricultural College. The time now is 6 o'clock on the Wednesday evening. Say what has already happened and what has not happened yet.

TIMETABLE

	MONDAY	TUESDAY	WEDNESDAY	THURSDAY	FRIDAY
09.00	Mr Smith (L) Livestock	Mr Smith (T) Livestock	Mr Rose (L) Soil	Mr Rose (D) Soil testing	Mr Roberts (L) Farm machinery
10.00	Mr Penn (L) Cereals	Mr Hamer (T) Vegetables	Mr Roberts (D) Tractors	Mr Rose (L) Irrigation	Mr Lever (L) Grain storage
11.00	Mr Roberts (D) Harvesters	Mr Hamer (L) Cabbages	Mr Smith (L) Cattle	Mr Penn (T) Rice	Mr Rose (T) Irrigation
14.00		Mr Hamer (T) Brassicas	Mr Smith (T) Cattle diseases	Mr Penn (T) Wheat	Mr Roberts (D) Ploughing

L = Lecture; D = Demonstration; T = Tutorial

Do you remember?
Did you go to Mr Smith's lecture *last Monday*?
No, I *missed* it. I *was* sick. I *fell* ill a week *ago*.

but

I *haven't seen* you *for three days*.
No, *I've been* busy. *I've had* two essays to write. *I've already finished* one, but I *haven't finished* the second yet.

2. Now complete the conversation between two students below. Be careful which tense you use!

Alan: I (not see) _____ you at Mr Smith's lecture yesterday, John. Where (be) _____ you?

John: I (forget) _____ all about it. I (start) _____ the essay two days ago and I (want) _____ to finish it before the lecture. It's a pity. It's the first lecture I (miss) _____ so far this term.

Alan: I (not complete) _____ the reading for that essay yet. I (go) _____ to the library yesterday, but the books (not be) _____ there.

John: I (never find) _____ the books I need in the library since I (come) _____ to this college a year ago. Someone (just take) _____ them out each time I go there. Something ought to be done about . . . .

E. Understanding a lecture

1. You are now going to hear part of a lecture, divided into short sections to help you understand it. As you listen, answer the questions below.

Section 1

- What is the subject of the lecture?
- What kind of animal is the lecturer going to talk about?
- What two examples of this kind of animal does the lecturer give?

Section 2

- What does the lecturer say is 'rather like a tube'?
- Is this statement correct or incorrect? ☐
  During digestion, complex and insoluble food parts are changed into simple and soluble ones.

Section 3

- Note down the three changes the lecturer tells you about
  ______________ are converted into ______________
  ______________ are converted into ______________
  ______________ are converted into ______________

Section 4

- Is this statement correct or incorrect?
  Cows swallow a lot of saliva with their food. ☐
- Label the diagram in the way the lecturer tells you to.

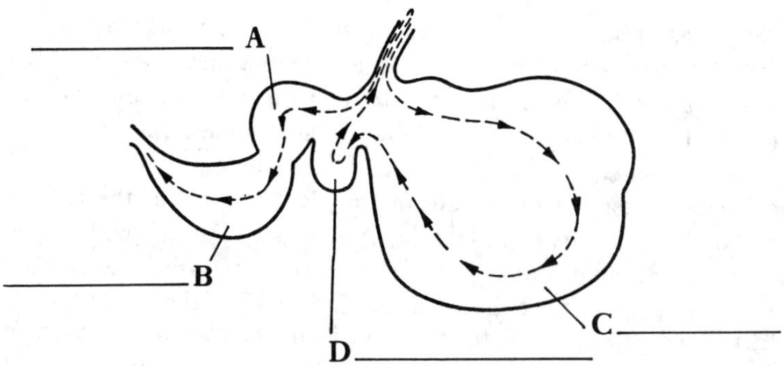

Section 5

- Label the diagram above in the way the lecturer tells you to.

Section 6

- Label the diagram above in the way the lecturer tells you to.

Section 7

- Label the diagram above in the way the lecturer tells you to.
- Look at the diagram on the next page. Note down which two parts of the digestive system the lecturer mentions.

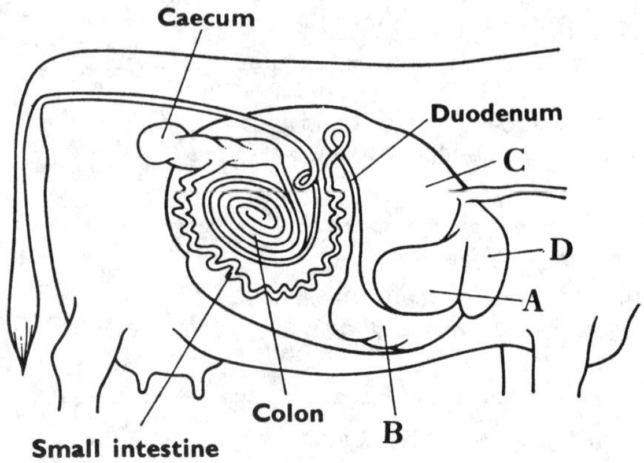

2. Now wind the tape back to the beginning of the lecture and listen to it again. This time, instead of answering questions, take notes. When you have listened to the whole of the lecture, you will be asked to provide a short summary of it. So make sure you note down the important points.

F. Understanding a printed text (2)

Read the following text carefully, looking up anything you do not understand.

The absorption and transport of food

1 Absorption involves the passage of food materials from the digestive tract into the blood stream. The blood then transports the food round the body.

The small intestine

2 Most absorption takes place in the small intestine which is specially adapted for the purpose. In the cow it is 40m long and in the pig it is 15m long. Its inner surface is folded into finger-like projections called *villi* (Fig. 6) which greatly increase the surface area for absorption. Within the villi there are fine blood capillaries and lymphatic vessels. The amino acids and glucose enter the blood stream directly, but most of the fatty acids enter the lymphatic system. (The latter eventually joins up with the blood stream.) In the ruminant some fatty acids, produced from the breakdown of cellulose, are absorbed directly through the wall of the rumen.

3 The blood stream carries the food nutrients, together with oxygen from the lungs, to the parts of the body where they are required. The lymph is then able to pass through the walls of the blood capillaries and can bathe the actual tissues of the body. At the same time it carries food substances to these tissues.

This text was taken from BUCKETT, *Introduction to Livestock Husbandry*, Pergamon Press

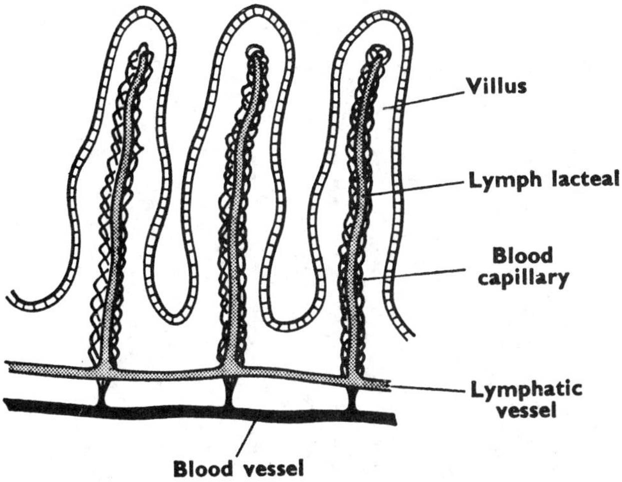

Fig 6. Structure of villi in small intestine

Digestibility of foodstuffs

4 Some of the food eaten by farm animals cannot be digested and passes out of the body as dung. The digestibility of the food is therefore very important and it is essential to consider it when feeding stock. Digestibility can be expressed as follows:

$$\frac{\text{Dry Matter eaten} - \text{Dry Matter in dung}}{\text{Dry Matter eaten}} \times 100.$$

The D-value

In practice farmers actually use the term D-value to compare the digestibility of foods. This is the percentage of digestible organic matter contained in the dry matter of the food. The higher the D-value the higher the digestibility.

5 This can be illustrated by reference to grass and silage. As grass gets older its cell walls get toughened with lignin and its digestibility falls. When making top quality silage farmers will aim to make a product which has a 70 D-value. This is very suitable for high yielding dairy cows but it can only be made from relatively young grass. As a result the yield per ha may be reduced compared to silage made from grass cut at a later stage. When feeding the dry, in-calf, beef cow some farmers consider quantity and bulk to be more important than quality. They cut the grass at a later stage and obtain silage with a D-value as low as 55 to 60, or even lower.

6 A low D-value will probably result in the slowing down of the food in its passage through the body. This will limit the quantity eaten by the animal. In high yielding dairy cows, or beef cattle required to make good liveweight gains, this can be a serious disadvantage. With both classes of stock it is essential that a large amount of food nutrients are digested and made available for use in the body. This is especially so with dairy cows in early lactation to avoid excessive loss of body weight and to promote high yields.

▼

G. Check your understanding

1. Can you explain the following words?

- capillary
- lymphatic vessel
- tissue
- silage
- liveweight
- lactation

2. Are these statements correct or incorrect?

- The small intestine carries food materials round the body.
- The lymphatic system never joins the blood stream.
- Both food nutrients and oxygen are carried by the blood stream.
- Farm animals digest all the food they eat.
- Food with a D-value of 65 is more digestible than a food with a D-value of 60.
- Silage with a D-value of 70 can be made from old grass.
- All farmers agree that quality silage is better than quantity.
- Animals eat less of a low D-value silage.
- A low D-value is better for dairy cows in early lactation.

H. Understanding discourse

A lecturer is giving the title for an essay to be written during the coming week. He is also giving the main points that must be covered. Listen to what he says and note down what you must do.

CHECK YOUR PROGRESS

1

The following short test is for you to check whether you are learning the skills you will need to study Agriculture in English. It is for *your* information. It is *not* an examination. You may use a dictionary.

A. Reading

Read the following text and then answer the questions.

ORIGINS OF CROP PLANTS

It is difficult, if not impossible, to pinpoint the exact origin of modern crop plants, in part because many of them were under cultivation long before the start of recorded history. It is also reasonable to assume that as the early agriculturists moved about, they carried plants with them far from their point of origin. Thanks to the selective processes of evolution, many of our crop plants are markedly different from their early ancestors. In fact, in the case of corn (maize), the earliest progenitors are thought to be extinct.

Despite the lack of solid evidence, the origin of many of our crop species is believed to be as shown in Fig. 1-1, the major centers of origin being southwestern and central Asia, the Mediterranean region, southeastern Asia, and the highlands of tropical America. Note that no major crops are shown as originating in the United States. Although not of major importance in world trade, the cranberry, blueberry, and pecan are native to the United States, as are certain species of plum, grape, and strawberry.

Although plant origins may seem to be only of academic interest, such information can be of real value. Plant breeders strive to improve crop plants by incorporating a wide range of genetic traits, e.g., disease or insect resistance, tolerance to adverse environmental factors, or modified growth habit. Often the best place to find genetic material with the desired characteristic is in the region to which the crop is native. There one can often find the broadest possible base of available genetic diversity in the relatives and progenitors of the crop species.

This text was taken from BARDEN *et al.*,
Plant Science, McGraw-Hill

PLANTS AS FOOD

Grains

Although the species vary with location and climate, the grain-yielding grasses are a major human food source in much of the world. In many temperate regions, for example, wheat is the primary grain staple (Fig. 1-2); in warmer, more humid areas, rice may fill this role. Other major grain crops in the grass family include corn, oats, rye, barley, millet, and grain sorghum. We discuss later how grains serve as a major feed source for animals destined for human consumption.

Starchy Foods

In addition to the cereal grains, starchy roots and other plant parts such as potato, sweet potato, cassava, yam, and taro (Fig. 1-3) have served as staple foods for centuries, especially in regions where grains are not well adapted. The potato has played a major role in world commerce and history. Cassava, also known as manioc, from which we get tapioca, has sustained people in the tropics for centuries. Although usually considered a fruit in the United States, the banana is a starchy staple in many tropical regions of the world.

Vegetables

Besides starchy products like the potato and sweet potato, many other vegetables contribute to the human diet. Favorite species vary with location in the world and even within the United States. Important vegetable crops include sweet corn, snap bean, pea, bean, lettuce, and tomato. These foods offer not only pleasing texture, flavor, color, and appearance but vitamins and minerals as well. Although most are not staple foods, they constitute vital portions of our diet.

Answer these questions.

1. The earliest cultivars of maize
- no longer exist ☐
- can be found in the wild ☐
- are still grown for food ☐
- are found both in the wild and under cultivation ☐

2. Is this statement correct or incorrect?
- The most important of modern American crops are native to the USA. ☐

3. In line 23, the words *such information* refer to
- plant breeders ☐
- academic interest ☐
- real value ☐
- plant origins ☐

4. Is this statement correct or incorrect?
- If a plant breeder wants to improve a crop plant he should never go back to the part of the world the plant originally came from. ☐

5. *Humid* means
- dry ☐
- damp ☐
- very wet ☐
- rainy ☐

6. Which of these crops are not grain-yielding grasses?

- millet ☐
- cassava ☐
- rye ☐
- taro ☐

7. Is this statement correct or incorrect?

- The sweet potato is a very recent source of food.

8. *Manioc* is another word for

- tapioca ☐
- potato ☐
- cassava ☐
- taro ☐

9. Is this statement correct or incorrect?

- People all over the United States like the same kind of vegetables.

10. In the last line but one, the word *they* refers to

- stale foods ☐
- vegetables ☐
- vitamins and minerals ☐
- vital portions ☐

B. Writing

Read the following text. Then, using your own words, write a short account of *The Organic Cycle*. You can refer back to previous Units in this book if you like.

Soil organic matter (O.M.)

This is anything previously living but now dead and in various stages of decomposition. Fully-decomposed O.M. is *humus*, a dark brown complex substance.

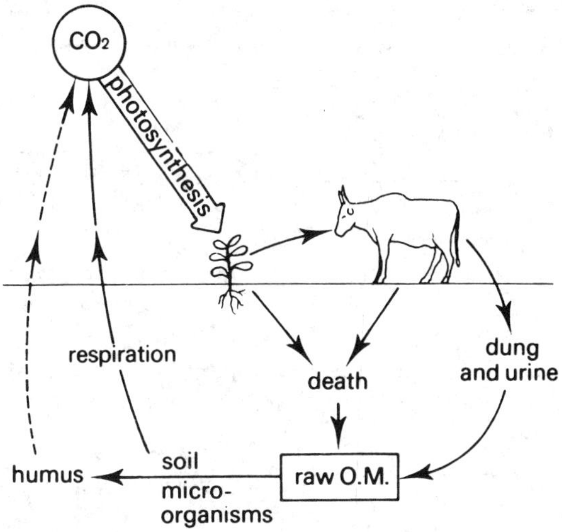

The organic cycle

Organic matter contains the principal reserve of nitrogenous compounds in soils. There is virtually no nitrogen (N) in rocks:

Decomposition of O.M.

(a) This is vital, since it removes dead matter, especially if the matter is diseased.
(b) It is also vital since it releases minerals to supply new life.
(c) It is rapid if soil is moist, warm and generally fertile.
(d) It is rapid if plant residues are young, thus rich in sugars, proteins and nitrogenous compounds, but low in fibre.

Humus

1. It holds moisture.
2. It holds *cations* (its capacity to do this is some fifty times greater than that of kaolinite clay).
3. It helps the aggregation of soil particles (structure).
4. It contains about 58% carbon; 5·8% nitrogen. Low-nitrogen materials use up extra soil nitrogen to decompose, e.g. cereal residues.

C. Listening

You will now hear part of a lecture. After each section, answer the questions below.

Section 1

- Is this statement correct or incorrect?
- The title of the lecture is Soil Erosion. ☐
- What are the two main agents of erosion?

____

Section 2

- How much topsoil can be lost through erosion every year?

________ per ________ per ________

- Note down where this loss can take place

____

____

- To how much does tree planting reduce this loss?

____

Section 3

- How many factors are there in soil erosion?

____

- Complete this statement correctly. Serious erosion takes place when the slope
  - is less than one degree. ☐
  - is between one and six degrees. ☐
  - is six degrees or more. ☐
  - is ten degrees or more. ☐

Section 4

- Note down why vegetation helps prevent erosion.

____

Section 5

- Are these statements correct or incorrect? The amount of humus in the soil makes no difference to erosion. ☐
- Is this statement correct or incorrect? Loams erode more quickly than dry, light soils. ☐

Section 6

- Note down the two factors in soil erosion mentioned by the lecturer.

____

- At what rate does rainfall become dangerous?

____

The passaages for Writing and Listening were taken from JOY & WIBBERLEY, *A Tropical Agriculture Handbook*, Cassell

UNIT 6

LIVESTOCK HEALTH

A. Understanding a printed text (1)

The following text will introduce you to the topic of **livestock health**. Look at the way it is divided into sections and paragraphs. Pay attention to the headings and notes in the margins.

Now look at these questions:

1. What is the purpose of this text?
2. Which animals are mentioned in the text?
3. What is a symptom?
4. What two kinds of symptom are described?

Read the passage through and find the answers to the questions. Remember, you do not have to understand every word to answer the questions.

LIVESTOCK HEALTH

Topic 1. The Symptoms of Ill Health

The cost of ill health

1 Livestock diseases cost farmers millions of pounds annually. In addition to deaths they cause loss of production and frequently a loss of body condition. Unthrifty animals require more food and take longer to fatten than healthy stock.

2 The good stockman must be capable of detecting early symptoms of disease so that prompt action can be taken to reduce loss, and to prevent the disease from spreading.

Typical Symptoms of Disease

Temperature

3 1. *Variations in temperature* can be measured by inserting a clinical thermometer into the rectum of an animal for 30 sec. High temperatures are usually associated with the increased activity of the body in fighting off disease. Young animals, females in late pregnancy and excited stock frequently have higher temperatures than the normal figures given below.

Cattle	38.6°C	(Range acceptable 37.8–39.2°C)
Sheep	39.4°C	(Range acceptable 38.9–40.0°C)

Pulse

4 2. *Variations in pulse* reflect the rate at which the heart pumps blood through the body. They can be measured with the index finger where arteries pass near the surface of the body. For example, in cattle the pulse can be taken under the tail. Normal pulse rates:

Cattle	50–60 beats per minute
Sheep	70–90 beats per minute

Other symptoms

5 3. *Rapid and irregular breathing* occurs in fevered conditions. Serious lung complaints may cause the animal to grunt with pain as it breathes.

This text was taken from BUCKETT, *Introduction to Livestock Husbandry*, Pergamon Press

6 4. *Loss of appetite* is common to many diseases, but mouldy or unpalatable food may be responsible.

7 5. *The coat* should not have bald patches. These usually indicate rubbing to relieve irritation caused by parasites such as lice. The condition of the coat will vary with housing conditions and grooming, but when cattle are infected with worms, or have wasting diseases, their coat loses its "bloom" or lustre. In sheep the coat must not be dull and show signs of falling off.

8 6. . . . . . . A tight *skin* over the ribs of cattle is particularly undesirable and is common to several ailments.

9 7. *The dung and urine* are stained or darkened if they contain blood. Scouring is common to many diseases, but this may occur when an animal is fed on a very laxative diet.

10 8. *Milk yield* in dairy cows will fall even if the cows have only a slight chill. Blood and clots within the milk indicate a disease of the udder; probably mastitis.

11 9. *The behaviour* of stock with certain diseases is abnormal. Animals with foot and mouth disease paddle their feet and produce large quantities of saliva, because they have sores on their feet and in their mouths. Cattle with grass staggers become excited and go into fits.

12 10. *The eyes* in healthy animals are bright and alert. When they are injured by pieces of straw, they become watery and produce a discharge. Animals with fevers usually have sunken eyes.

13 11. *The head* of a healthy animal is usually held fairly upright. Sick stock hang their head low and have a dejected appearance.

The stockman's responsibility

14 The combination of symptoms shown by an animal will help to diagnose the disease from which it is suffering. The responsibility of the stockman is to detect the early symptoms. In some cases he will be able to treat the animal himself, e.g. by drenching. However, the prompt action of a veterinary surgeon, who has training and laboratory facilities behind him, may be necessary in many cases.

Clinical and sub-clinical symptoms

15 Symptoms which can be seen by examination of the live animal on the farm are called *clinical symptoms.* Animals can have mild forms of a disease which do not produce sufficient symptoms for the stockman to see. Such conditions of disease are said to be *sub-clinical.*

B. Check your understanding

Now answer these questions.

1. Why should a stockman be able to recognise symptoms?
2. What temperature should cattle normally have?
3. What is the normal pulse rate of a sheep?
4. What symptom does an animal with a serious lung complaint show?
5. What may be the cause of an animal not eating properly?
6. What are the possible causes of a poor coat?

▼

7. What are the symptoms of mastitis?

8. How can you tell foot and mouth disease?

9. How does a healthy animal hold its head?

10. When should the stockman call the vet?

C. Increase your vocabulary

In this section you should use your dictionary to help you answer the questions about the text.

1. Look at paragraphs 1 and 2 again. What words have the same meaning as:

- be able to
- noticing; finding out
- are the reason for
- quick and immediate
- need

2. Look at paragraph 3 again. What words have the opposite meaning to:

- taking out
- calm

3. Now look at paragraphs 4 and 5 again. Can you explain the words:

- arteries
- fevered conditions
- lung complaints

4. Look at paragraphs 6 and 7 again. Which words correspond to the definitions:

- an animal which lives on another
- having a woolly or furry growth of fungi
- looking after by feeding and brushing
- a condition that makes animals scratch themselves

5. Look at paragraphs 8 to 10 again. What words have the opposite meaning to:

- great, large
- lightened
- loose

6. Look at paragraphs 11 and 12 again. Can you explain the words:

- saliva
- sores
- staggers
- discharge

7. Look at paragraphs 13 to 15 again. What words have the same meaning as:

- enough
- not very bad/serious
- in low spirits

D. Check your grammar

MAKING COMPARISONS

Do you remember?
An N'dama bull has *longer* horns than a Shuwa Arab.
The Kuri bull has the *longest* horns.
The Shuwa Arab's horns are *not as long as* the N'dama's.

1. Look at the pictures of the four different breeds. Make comparisons between the breeds using the words in the first picture.

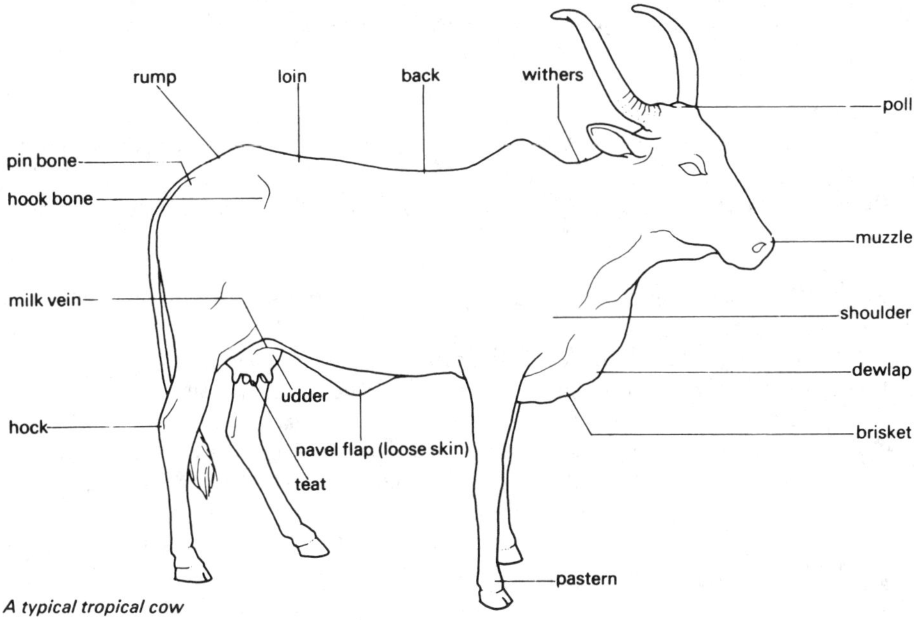

A typical tropical cow

An N'dama bull

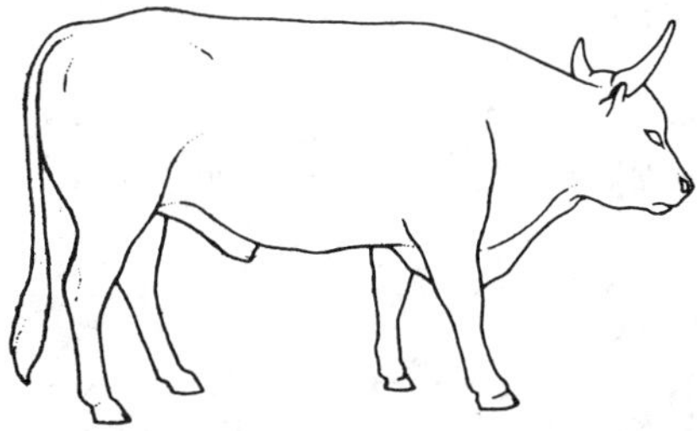

A Kuri bull

Mutura

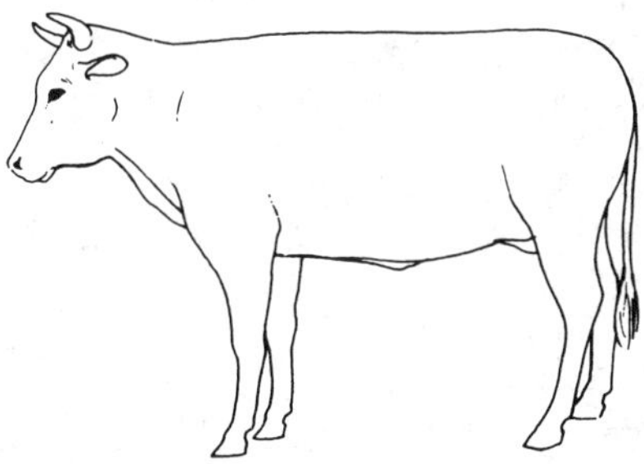

Shuwa Arab

Drawings from JOY & WIBBERLEY, *Tropical Agriculture Handbook*, Cassell

2. Join the following pairs of sentences together to make single sentences by using either *although* or *whereas*.

Example:
The good stockman must recognise the symptoms of disease. He may not be able to cure it himself.
The good stockman must recognise the symptoms of disease, *although* he may not be able to cure it himself.

- The N'dama has an average liveweight of 250–350 kg. The Kuri has a mature liveweight of 450–500 kg.
- The Shuwa Arab is usually dark red. This may sometimes be mingled with black.
- The Mutura fatten well. They are often only slaughtered on ceremonial occasions.
- The N'dama are dun or fawn in colour. The Mutura are usually black.
- The Shuwa Arab is mainly used for milk production. The Muturu are rarely milked.
- The Shuwa Arab is found in N.E. Nigeria. The Kuri inhabit the islands of Lake Chad.

E. Understanding a lecture

1. You are now going to hear part of a lecture, divided into short sections to help you understand it. As you listen, answer the questions below.

Section 1

- What is the subject of the lecture?
- Note down the two commonest shapes of bacteria.

____

- Which kind are called 'cocci'?

____

Section 2

- Note down the name and characteristics of the bacteria the lecturer talks about.

- ______________________________

____

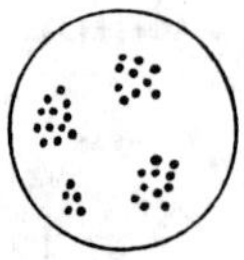

- ______________________________

____

Section 3

Note down the name and characteristics of the bacteria the lecturer talks about.

- ______________________________

____

- ______________________________

____

Section 4

Are these statements correct or incorrect?

- All bacteria produce spores. ☐
- The anthrax bacillus can survive in adverse conditions. ☐

Section 5

Are these statements correct or incorrect?

- Bacteria prefer a temperature of more than 20°C. ☐
- All bacteria need moisture. ☐
- All bacteria need air. ☐

Section 6

Complete the statement correctly.

- Disinfectant is effective when
  there is no manure in the building. ☐
  there is a little manure in the building. ☐
  there is a lot of manure in the building. ☐

2. Now wind the tape back to the beginning of the lecture and listen to it again. This time, instead of answering questions, take notes. When you have listened to the whole of the lecture, you will be asked to provide a short summary of it. So make sure you note down the important points.

F. Understanding a printed text (2)

Read the following text carefully, looking up anything you do not understand.

Contagious and infectious diseases

Topic 2: The causes of ill health

Spread of Disease

1 Bacterial diseases are *contagious*. This means that they can be passed from one animal to another. Some bacterial diseases can only be spread by contact between stock and may be controlled by isolation. Others are highly *infectious*, and can be spread on stockmen's clothes, in food and on lorry wheels, etc.

2 Bacteria enter the body through the mouth, nose, eyes and other body orifices. Wounds and breaks in the skin are also common paths of entry. Infection may be only local, that is, confined to the site of access, or it may be carried all over the body.

3 Viruses are even smaller than bacteria and are very infectious. A warm humid building is particularly conducive to the spread of virus diseases.

4 Foot and mouth disease is a typical example of a virus disease and the ease with which it spreads is well known.

This text was taken from BUCKETT, *Introduction to Livestock Husbandry*, Pergamon Press

Toxic fluids

Immunity

5 Certain bacteria and viruses cause disease because fluids which they produce are poisonous, or *toxic*, to stock. The animal body has several defensive mechanisms against them. The most important of these are substances called ***antibodies,*** which either destroy the disease organisms or neutralise the effect of their *toxins*.

6 There are several ways in which an animal can get *immunity*. Some are natural, others are artificial, or man-made.

Natural immunity

7 *Natural immunity*. When an animal contracts a disease it produces antibodies against it. If the defensive mechanisms of the body are sufficient the animal will recover. However, the body will usually continue to produce antibodies against that disease for several years, or throughout its life. The body has, therefore, acquired immunity naturally. It is prepared for a subsequent attack and, unless it becomes heavily infected, it will destroy the disease organisms before they can multiply and produce symptoms of the disease.

Antibodies

8 Unfortunately, antibodies are fairly specific. This means that the antibodies against one disease will not attack organisms which cause a different disease.

9 Occasionally animals occur which have developed a natural immunity and can carry the infective organisms in their body, without destroying them. Human *carriers* of typhoid bacteria are particularly dangerous, especially if they have contact with food, because they may transmit the disease to others.

The importance of the mother's milk

10 Mothers are capable of passing on some immunity to their offspring, either while they are pregnant or, more particularly, in their first milk, the *colostrum*. This is why it is essential to see that all young animals get their mother's milk within the first day of life. It is particularly important to sheep farmers who produce an artificial immunity in the ewe against lamb dysentery, which she passes on to the lambs in her milk.

11 Certain species are immune naturally to diseases which attack other species. For example, horses and poultry do not suffer from foot and mouth disease.

Artificial and active immunity

12 *Artificial immunity*. Artificial immunity can be induced in animals (Fig. 8), by inoculating them with *vaccines*. These are fluids which contain bacteria or viruses in suspension that have been killed, or modified to slow down their activity. The body then produces antibodies against them and may develop an *active immunity*, lasting for several years, against the particular disease which they cause.

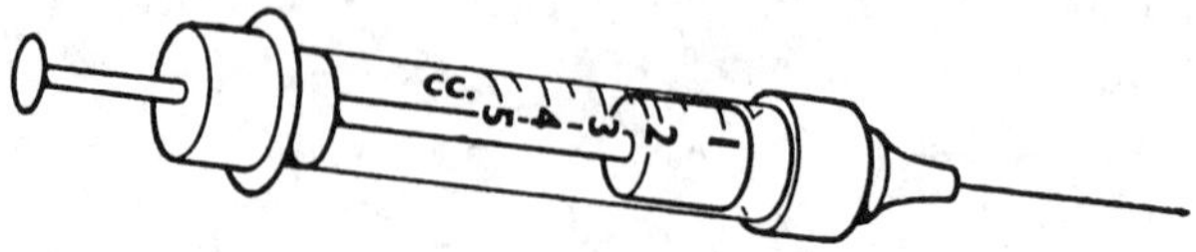

FIG. 8. Syringe for injections.

When to inject serum

13 If an animal already has a disease vaccines must not be used. They take several days to produce immunity and inoculation at this time may even add to the illness. However, an injection with a *serum*, which already contains antibodies against the disease, will prove effective. It does not stimulate the body to produce antibodies itself, because no disease organisms are actually introduced. It therefore confers a *passive immunity* which only lasts a short time.

14 Injections of sera are not always restricted to animals with disease. For example, in sheep, lamb dysentery kills many lambs in the first 2 or 3 weeks of life. If their mothers were not vaccinated it may be wise to inject all lambs on farms where the disease is common, with sera. Vaccination of the lambs would not produce immunity in time to prevent loss.

G. Check your understanding

1. Look at paragraphs 1 to 4. Can you explain the difference between
- a contagious and an infectious disease
- bacteria and viruses

2. Look at paragraphs 5 to 8 again. Complete these statements correctly.
- Some bacteria bring disease because
  - they produce toxic fluids. ☐
  - they contain antibodies. ☐
  - they destroy the animal's antibodies. ☐
  - they attack the animal's fluids. ☐
- An animal acquires natural immunity when
  - it is injected with antibodies. ☐
  - it is given artificial antibodies. ☐
  - its own body produces antibodies. ☐
  - it has recovered from a disease. ☐
- Antibodies
  - will attack any disease the animal gets. ☐
  - will not attack any disease the animal gets. ☐
  - are directed against a particular disease. ☐
  - only defend the animal when it is heavily infected. ☐

3. Look at paragraphs 9 to 11 again. Explain in your own words why a mother's milk is important.

4. Look at paragraphs 12 to 14 again. Are these statements correct or incorrect?
- An active immunity will last the whole of an animal's life. ☐
- Vaccines do not contain viruses. ☐
- You use a vaccine when an animal already has a disease. ☐
- You use a serum when an animal already has a disease. ☐
- You should inject lambs with sera if their mothers were not vaccinated. ☐

H. Understanding discourse

Listen to the lecturer telling you the procedure to follow if you want to use a laboratory on your own. Note down the procedure you must follow.

UNIT 7

VEGETABLES (1)

A. Understanding a printed text (1)

The following text will introduce you to the topic of **vegetable growing** in the United States. Look at the way it is divided into paragraphs. Pay attention to the notes in the margins.

Now look at these questions:

1. How is the writer going to classify vegetables?
2. Does the writer describe the kind of soil required for vegetable growing?
3. Are any vegetables hand-harvested in the USA?
4. Does the writer describe different cropping systems?
5. Is vegetable growing in your country different? If so, how?

Read the passage through and find the answers to the questions. Remember, you do not have to understand every word to answer the questions.

Vegetable growing

It is difficult to classify vegetables

1 The wide diversity of vegetables makes a generally acceptable system of classification hard to devise. We shall use a combination of criteria—some genetic, some morphological, and some based on the plant part eaten. Terminology is further confused by the lack of clear distinction between fruits and vegetables. A *vegetable* can be defined as the edible portion of herbaceous garden plants, usually grown as annuals. With lettuce, celery, potato, and carrot, the edible portion is the leaf, petiole, tuber, and root, respectively. With tomato, melons, and cucumber, however, it is the fruit that is eaten. Most *fruit* plants are perennial, some woody, others herbaceous. A further distinction between fruits and vegetables is that fruits are generally eaten as a dessert or snack, a vegetable as a part of a main course.

A vegetable

A fruit

General principles of vegetable production

2 Although the production systems used for vegetable crops vary with region, crop, etc., some general principles apply. For successful vegetable production, the soil should be of good tilth, well drained, and deep enough to accommodate the root system. Soil tests, standard procedure to determine the need for fertilizer or pH adjustment, should be made before soil preparation so that supplements can be incorporated into the field as it is plowed or otherwise tilled.

The seedbed

3 A basic requirement in establishing a vegetable crop is a relatively smooth, clodfree seedbed, prepared by plowing, disking, and sometimes further smoothing of the soil surface. A very fine seedbed is much more critical for small-seeded species, e.g., lettuce, than for large-seeded crops, e.g., snap bean and sweet corn. In areas where furrow irrigation is used, a system of ridges and furrows is formed before seeding the crop.

4 Vegetables are grown in rows to simplify the operations involved from seeding to harvest. With small species, e.g., radish and onion, the rows are quite close together (38 to 40 cm); with large ones, e.g., cucumber, rows are usually 1.2 to 1.8 m apart. Within-row spacing ranges from 2 cm for radish to 1 m or more for cucumber. The vast majority of vegetable crops are seeded directly in the field with various types of planters. The development of precision planters which space seed at even intervals has reduced the need for subsequent thinning. For early marketing of such high-value crops as tomato, pepper, and cabbage, transplants are set in the field. The transplants are purchased from southern areas or are grown in greenhouses, cold frames, etc. The modified planting techniques for potato, sweet potato, and cassava will be described later in the chapter.

Vegetables are grown in rows

5 The vegetable industry has undergone major changes to reduce labor inputs. Herbicides have largely replaced hand hoeing and cultivation. Mechanical harvesters are standard for many of the major vegetables, particularly those destined for processing markets. However, lettuce, cantaloupe, and watermelon, are still largely hand-harvested because judgment is needed to evaluate the maturity of the product.

Changes in the vegetable industry

6 The need for irrigation is obvious in arid regions, but even in humid areas, irrigation is beneficial in ensuring rapid seed germination, seedling survival, and optimal product quality and yield.

Irrigation

7 The length of the growing season, climate, and types of vegetables grown determine the cropping system. Since radish, spinach, and snap bean mature quickly, successive plantings can be made. In other situations, a cool-season crop, e.g., spinach, may be followed by later plantings of warm-season species, e.g., snap bean, which in turn may be followed by a fall crop of cabbage. Many multiple-cropping combinations are possible and are selected largely on the basis of the anticipated market. Long-season crops, e.g., potato in the north, may occupy most of the growing season. A cover crop is often used to minimize erosion and build the soil during the off-season.

Cropping systems

This passage was taken from BARDEN *et al.*, *Plant Science*, McGraw-Hill

B. Check your understanding

Now answer these questions.

1. Which part of celery is eaten?
2. How are vegetables eaten in the USA?
3. How deep should the soil be for vegetable growing?
4. When should soil tests be made?
5. For what kinds of vegetable is a fine seedbed required?
6. Why are vegetables grown in rows?
7. Why are precision planters an advantage?
8. Why is lettuce still hand-harvested?
9. Why is irrigation useful in humid regions?
10. Why is a cover crop often used?

C. Increase your vocabulary

In this section you should use your dictionary to help you answer the questions about the text.

1. Look at the first paragraph again and say which words match these definitions.

- having stems that are not woody
- a plant that lasts from year to year
- can be eaten
- enlarged part of an underground stem

2. Now look at paragraph 2 again. What words have the same meaning as:

- to find out; to establish
- to provide room for
- way of doing things
- put into

3. Now look at paragraph 3 again. Can you explain the following words:

- clodfree
- disking
- furrow irrigation

4. Now look at paragraph 4 again. What words have the opposite meaning to:

- far apart
- make different
- very small
- preceding

5. Now look at paragraph 5 again. Can you explain the following words:

- herbicide
- hoeing
- processing market
- maturity

6. Now look at paragraphs 6 and 7 again. What words have the same meaning as:

- take up
- chosen
- one after the other
- because
- autumn
- expected

D. Check your grammar

ASKING QUESTIONS

Can you ask questions in a way that makes sure you get the information you want? And can you understand the answers? You have to write a short report on the potato in the USA. The teacher has the information you need to do this. Ask the questions you need to ask and note down the answers. You should ask about:

- Latin name
- importance of the potato
- growing conditions
- the plant itself
- cultivation
- harvesting
- storage

E. Understanding a lecture

1. You are now going to hear part of a lecture, divided into short sections to help you understand it. As you listen, answer the questions below.

Section 1

- Which group is the tomato a member of?

____

- Note down two other vegetables in this group.

____

- Is this statement correct or incorrect? The potato and tomato are members of the same family. ☐

Section 2

- Complete the statement correctly.
  The tomato originated in
  Mexico
  Europe
  United States
  South America

Section 3

Are these statements correct or incorrect?

- The tomato is grown as an annual all over the world. ☐
- Greenhouses are often used for early tomato production in northern Europe. ☐

Section 4

- Note down the distinction the lecturer makes.

____

Section 5

- What term does the lecturer use for 'different kinds' of tomato?

____

- How long are most common tomato stems?

____

- How many fruits are there in a cluster?

____

Section 6

Are these statements correct or incorrect?

- In ground production, the plants are pruned. ☐
- Ground production is used for tomatoes for processing. ☐
- For fresh tomato production the plants are usually supported. ☐
- Note down the three advantages of supporting and pruning tomato plants. ☐

____

____

Now wind the tape back to the beginning of the lecture and listen to it again. This time, instead of answering questions, take notes. When you have listened to the whole of the lecture, you will be asked to provide a short summary of it. So make sure you note down the important points.

F. Understanding a printed text (2)

Read the following text by an American writer carefully, looking up anything you do not understand.

Salad Crops

1 The salad crops include lettuce, celery, escarole, and a variety of miscellaneous green vegetables such as chicory and parsley. We shall consider only the first three, which rank among the more important vegetables.

Lettuce (Compositae)

2 Lettuce (*Lactuca sativa*) is in the sunflower family. It is widely accepted that *L. sativa* developed from *L. serriola*, the wild prickly form of lettuce native to Asia, Europe, and northern Africa. Lettuce was known and eaten by the early Greeks and Romans and even earlier by the Chinese.

The four main types

3 The four major types of lettuce, all classified as *L. sativa*, are crisphead; butterhead; cos, or romaine; and leaf, or bunching. In China, stem, or asparagus, lettuce is widely grown; the fleshy stem is eaten fresh or cooked and the leaves are discarded.

4 Most of the commercial lettuce crop in this country consists of the crisphead type, characterized by firm heads and brittle texture of the leaves. For short-distance shipment, the butterhead and romaine types of lettuce are of some importance. The leaf lettuce cultivars are produced mostly in home gardens and in limited quantities in greenhouses during the winter.

Use

5 Use of lettuce in this country is almost exclusively as a fresh vegetable, although in some parts of the world it is cooked like spinach. Lettuce is frequently the main constitutent of salads because of its attractive color, mild flavor, crisp texture, and low calorie content.

What lettuce is

6 Lettuce is a shallow-rooted annual plant which forms a rosette of leaves at the base and (if left unharvested) a flower stem which may be 30 to 100 cm tall. Some cultivars form a definite head whereas others are merely a loosely formed rosette of leaves.

7 Lettuce is propagated by seeds which are very small, averaging over 100 per gram. In the major lettuce-growing areas, direct seeding in the field is standard practice.

8 Much hand labor is involved in the commercial lettuce harvest. Workers walk through the lettuce fields and cut heads which are mature, as indicated by both size and firmness (Fig. 20-6). After the lettuce is packed in fiberboard boxes, water is sprinkled over it, the top of the box is stapled, and the cartons are hauled to a *vacuum cooler*; this large vacuum chamber holds from 320 cartons up to a loaded freight car. The pressure is lowered from the normal 2.57 MPa to about 0.015 MPa, at which pressure water "boils" at 0°C. As the water vaporizes, it absorbs tremendous quantities of heat, so that the lettuce can be taken from a 24°C field temperature to 0°C in about 30 min.

Harvesting lettuce

9 If lettuce is harvested, handled, and cooled properly, it will store at 0°C for about 3 weeks. Lettuce is normally available every day of the year to consumers in the United States.

Storage time

FIGURE 20-6
Lettuce is harvested by hand and packed in cardboard boxes in the field. It is then hauled from the field and placed in a vacuum cooler to remove field heat. (*U.S. Department of Agriculture.*)

Origins

Celery (Umbelliferae)

10 Celery (*Apium graveolens* var. *dulce*) is the second leading United States salad crop, with production about one-third that of lettuce. Celery apparently developed from a wild marsh plant occurring throughout temperate Asia and Europe. Early use of celery was as a medicinal herb; later it was used as flavoring, but it has been accepted as a food since the seventeenth century. Early forms of celery were not particularly tender or tasty and were used mainly in soups. Not until mild, tender cultivars were developed did it become popular as a fresh vegetable. Like lettuce, celery is available year-round to the North American consumer.

Bolting

11 Celery is normally a biennial plant, but certain stimuli can cause it to bolt. Ideal nonbolting temperatures are 15 to 18°C; temperatures above 21 to 24°C slow growth and lower quality; temperatures below 12 to 14°C for more than a few days may induce bolting.

The edible parts

12 The edible portions of the celery plant are the thickened petioles, or leaf stalks, which arise from a very compact stem. Celery is classified as an herb and grows to about 1 m. The dried fruits, often called seeds, although the actual seeds inside the fruits are minute, are used as flavoring, because they contain the same aromatic oils as leaves and petioles. Many other members of this family, e.g., anise, are used as herbs and spices.

Propagation

13 Celery is seed-propagated but with difficulty because the seeds are very small and germinate slowly. In California, considerable direct seeding is done, but in other areas most celery is started in greenhouses, cold frames, or protected beds and transplanted (Fig. 20-7).

Harvesting

14 Celery has been hand-harvested until recently, when mechanization has increased. After packing, the celery must be cooled quickly by icing, vacuum cooling, or hydrocooling, i.e., flooding the produce with water at 0 to 2°C.

Storage

15 Celery can be stored for up to a month if temperatures are at 0°C and the relative humidity is high. Polyethylene sleeves and shrink-film wraps are often used to provide mechanical protection and reduce water loss.

▼

FIGURE 20-7
Mechanical transplanters are widely used to set out young plants imported from southern areas or started in greenhouses. Some models apply water or a fertilizer solution to the young transplants. (Mechanical Transplanter Company)

This passage was taken from BARDEN *et al.*, *Plant Science*, McGraw-Hill

G. Check your understanding

1. Look at paragraphs 1 and 2 again. What words have the same meaning as:

- of mixed sorts
- originally belonging to
- most people agree

2. Look at paragraph 3 again. Complete the statement correctly.
The Chinese eat the stem of the

- asparagus lettuce ☐
- cos lettuce ☐
- butterhead lettuce ☐
- bunching lettuce ☐

3. Look at paragraph 4 again and note down two characteristics of crisphead lettuce.

____

4. Look at paragraphs 5 to 7 again. Complete the following table about the lettuce using the information in these paragraphs.

Flavour	
Texture	
Calorie value	
Length of life	
Root formation	
Leaf formation	
Method of propagation	
Weight of seeds	
Sowing method	

5. Look at paragraphs 8 and 9 again. What words have the same meaning as:

- shown
- pulled, dragged
- work
- keep
- take in liquid or light

6. Look at paragraphs 10 and 11 again. Which words correspond to the definitions:

- lead to or cause something
- run to seed
- area of low-lying, wet land
- a plant used in medicine or for flavouring food
- delicate; easily damaged

7. Look at paragraphs 12 to 15 again. Can you explain the words:

- aromatic
- spice
- germinate
- cold frame
- shrink-film wraps

H. Understanding discourse

Listen to a lecturer giving instructions to some students. Answer the questions below. Note that the lecturer does not give all the information.

1. What is the subject of next week's lectures?

____

2. The two books to be read are:

Title ______________________________

Author(s) ______________________________

Chapter(s) ______________________________

Pages ______________________________

Title ______________________________

Author(s) ______________________________

Chapter(s) ______________________________

Pages ______________________________

3. The article to be read is:

Journal ______________________________

Volume ______________________________

Issue ______________________________

Author ______________________________

Title ______________________________

4. What must the students find out before next week?

____

____

____

UNIT 8

FRUITS AND BEVERAGES

A. Understanding a printed text (1)

The following text will introduce you to the topic of **fruits**. Look at the way it is divided into sections and paragraphs. Pay attention to the headings and notes in the margins, and to the illustrations and captions.

Now look at these questions:

1. How many categories of fruit does the writer give?
2. Does the writer describe fruit propagation?
3. What type of fruit does the writer not exemplify?
4. Which parts of the world produce both pineapples and bananas?
5. Are pineapples and bananas harvested at the same stage of ripeness?

Read the passage through and find the answers to the questions. Remember, you do not have to understand every word to answer the questions.

Fruits

1 From a botanical standpoint, a fruit is an expanded and ripened ovary with attached and subtending reproductive structures. As such, fruits include a range of products from apples and berries to beans, peas, peppers, tomatoes, and squash. More commonly a fruit is considered as a fleshy, usually sweet, plant product eaten out of hand, in salads, or as desserts.

Classification of fruits

2 Although fruit-producing plants can be classified in many ways, one of the most important distinctions can be made on the basis of climatic requirements:

3 *Temperate fruits* are deciduous and tolerate or even require a cool period each year.

4 *Tropical fruits* are evergreen, can withstand no temperatures below freezing, and may, like banana, suffer chilling injury at temperatures somewhat above freezing.

5 *Subtropical fruits* are intermediate between the temperate and tropical types, some being deciduous and others evergreen. They can ordinarily withstand temperatures slightly below freezing but are seriously injured by temperatures of −4 to −5°C.

Most fruits are propagated vegetatively

6 With a few exceptions, the major fruit and nut crops are propagated vegetatively because they are open pollinating and do not breed true (Chap. 14). A seed from a 'Golden Delicious' apple will produce an apple tree, but its fruits will not resemble a 'Golden Delicious.' Although the seeds from a 'Bartlett' pear, a 'Redhaven' peach, a 'Parson Brown' orange, or a 'Hartley' walnut will produce a tree of the respective species, the fruits will in all likelihood be of very inferior quality. The same is true for the small fruits—strawberry, blueberry, grape, and cranberry.

7 Since the tree fruits, nuts, and grapes are budded or grafted on various types of rootstocks, the thousands or millions of plants of a particular scion cultivar of apple, orange, or almond have the same genetic makeup as the original plant (see Chap. 15). The rootstocks may be of seedling origin or be vegetatively propagated for a particular characteristic such as dwarfing, pest resistance, or tolerance to a specific climatic or soil factor.

Rootstocks

Tropical and subtropical fruits

8 Pineapple (Bromeliaceae)

The pineapple (*Ananas comosus*) is a low-growing perennial of the tropics. Although severely injured by temperatures of 0°C, the pineapple does not thrive at temperatures above 30 to 32°C. Features that make the plant quite resistant to desiccation include crassulacean acid metabolism, sunken stomata, narrow leaves, and funnel-shaped leaf bases which collect water.

Structure and development of the pineapple

9 The pineapple fruit is complex both in structure and method of development. The fruit is the result of about 100 individual flowers in an inflorescence at the apex of a vegetative shoot. Distal to the fruit a crown of small leaves grows. The fruit consists of the combined mass of the many individual fruitlets surrounding the inner stem tissue. When the fruit is prepared for use, the leathery outer rind is removed, as is the fibrous stem tissue in the center. Canned pineapple rings are the result of trimming the outer rind and removing the interior stem.

10 World pineapple production is in tropical regions such as South America, Asia, Africa, and Central America. Hawaii was the leading producer of pineapple, but much of the industry has shifted to areas where labor costs are lower.

Method of growing

11 Pineapples are grown in solid plantings of one cultivar to avoid cross pollination, which would result in seed formation. Fortunately, the fruits develop parthenocarpically.

Propagation

The three-crop cycle

12 Since the pineapple is seedless, it is propagated vegetatively by *slips* which arise beneath the fruit, *suckers*, which are shoots originating lower on the stem, or most commonly *crowns* at the top of the fruit (Fig. 21-16). To harvest pineapples for processing, workers wearing heavy chaps and gloves walk between the rows, harvest the fruit, and break off the crown. A week or two later the crowns are collected from the field and planted, often in fumigated soil using black plastic mulch. Newly set crowns normally take 18 to 24 months to flower, a process sometimes accelerated and commonly synchronized by the use of ethylene. The first crop consists of a single large fruit on each plant; the second crop, or first ratoon crop, consists of two smaller fruits per plant and is mature about 12 months after the first; the third crop, or second ratoon crop, consisting of four smaller fruits, is harvested about 12 months after the second crop. After this crop has been harvested, the field is prepared for replanting and the three-crop cycle repeated.

▼

Harvesting

13 Most pineapples are harvested when they are fully ripe and processed immediately. The fruits ripen after harvest but since they lack starch reserves get no sweeter. Modern jet airplanes transport increasing quantities of fresh pineapples to the United States and Europe. Even these are harvested before becoming fully ripe on the plant and therefore are inferior to those available locally in pineapple-producing areas.

14 Banana (Musaceae)

The banana (*Musa* spp.) is a tropical fruit which is productive between temperature extremes of 10 and 40°C but is best suited to a narrower range of 15 and 35°C. The starchy cooking banana, or *plantain* (*M. paradisiaca*) is common in the tropics, but the table cultivars in world trade are *M. sapientum*. The only state that produces bananas commercially is Hawaii. Most bananas sold in American markets are imported from Central and South America. Leading countries in banana production are Brazil, Ecuador, Costa Rica, Mexico, and several African nations.

The development of the banana

15 The banana is a fast-growing, herbaceous, monocotyledonous plant which develops suckers (pseudo stems) from underground rhizomes and reaches a height of 1.5 to 6 m. The leaves are large, up to 3.5 m long; the pseudo stem is composed of compressed leaf sheaths. After growing for several months, the flower stalk emerges from the top and usually bends downward (Fig. 21-17). Female flowers produce bunches, or *hands*, of bananas parthenocarpically. Since a sucker fruits only once, it is cut back to the ground after harvest. The underground rhizome is perennial, and new suckers develop while the one or two mature suckers are fruiting. New plantings are established by cutting and planting young suckers, 0.3 to 1 m tall, with part of the rhizome attached. Suckers develop rapidly, and the first harvest begins about 9 months after planting.

Harvesting

16 Bananas are harvested when mature but hard and green and develop excellent quality when ripened. Ripening is done at the distribution center by exposing the fruit to 1000 ppm ethylene for 24 h at 14 to 16°C. The ethylene treatment triggers natural ripening and also makes a shipment of bananas ripen more uniformly. Bananas are subject to chilling injury at temperatures below 10°C and so are not held in cold storage like most fresh fruits and vegetables.

This passage was taken from BARDEN *et al.*, *Plant Science*, McGraw-Hill

B. Check your understanding

Now read the text carefully, looking up any new items in a dictionary or reference book. Then answer the following questions:

1. What kinds of fruit require a cool period?
2. Which kinds of fruit are both deciduous and evergreen?
3. Why are most fruit crops propagated vegetatively?
4. What effect does a temperature of 0°C have on pineapples?
5. How are pineapples grown?
6. Why is ethylene used in growing pineapples?
7. When are pineapples harvested?
8. What temperature suits bananas best?
9. How are new plantings of bananas made?
10. When are bananas harvested?

C. Increase your vocabulary

In this section you should use your dictionary to help you answer the questions about the text.

1. Look at paragraphs 1–5 again. What words have the same meaning as:

- made larger
- needs
- making cold
- of the structure of plants
- hurt

2. Look at paragraph 6 again. Say what these words refer to:

- line 2: they
- line 5: its

3. Look at paragraph 7 again. Can you explain the words:

- budded
- grafted
- rootstock
- dwarfing

4. Look at paragraph 8 again and say which words match these definitions:

- grow strong and healthy
- process or state of drying out
- the biochemical processes an organism must accomplish in order to live
- a microscopic opening in leaves.

5. Look at paragraph 9 again. What words have the opposite meaning to:

- bottom or lowest point
- taken apart
- putting

6. Look at paragraphs 10–12 again. Can you explain the words:

- cross pollination
- fumigated
- mulch
- three-crop cycle

▼

7. Look at paragraph 13 again. What words have the same meaning as:

- do not have
- not as good as

8. Look at paragraphs 14–16. What words have the same meaning as:

- brought in from abroad
- pressed together
- comes out
- false
- is the cause of

D. Check your grammar

DEFINING

Do you remember?
A farmer is a person *who* grows crops.
The banana is a plant *which* develops suckers.

Define each of the words given below:

- Temperate fruits are fruits . . .

  ____
- A rhizome

  ____
- A rootstock

  ____
- Salads

  ____
- A sucker

  ____
- A slip

  ____
- An importer

  ____
- A monocotyledonous plant

  ____
- Herbicides

  ____
- An agricultural worker

  ____

E. Understanding a lecture

1. You are now going to hear part of a lecture, divided into short sections to help you understand it. As you listen, answer the questions below.

Section 1

- Note down two countries which import mangoes.

  ____

- How high does a mango tree grow?

  ____

- Is this statement correct or incorrect?
  Mangoes are only grown in India. ☐

Section 2

- Note down a definition of inflorescence.

  ____

- Note down a definition of panicle.

  ____

- Is this statement correct or incorrect?
  Mango trees develop hundreds of fruits. ☐

Section 3

- Note down the size of the fruit.

  ____

Section 4

- Are these statements correct or incorrect?
- A drupe has a thin exocarp. ☐
- A peach is classified as a drupe. ☐

Section 5

- Complete the following statement correctly.
  You harvest mangoes when
  the flesh is green. ☐
  the flesh has softened. ☐
  the flesh turns from green to yellow. ☐
  the fruit is fully ripe. ☐

2. Now wind the tape back to the beginning of the lecture and listen to it again. This time, instead of answering questions, take notes. When you have listened to the whole of the lecture, you will be asked to provide a short summary of it. So make sure you note down the important points.

F. Understanding a printed text (2)

Read the following text carefully, looking up anything you do not understand.

BEVERAGE CROPS

1 The main nonalcoholic beverages of world trade are coffee, tea, and cocoa, all grown in the tropics. In the United States, we drink much more coffee than tea, whereas the reverse is true in the world as a whole. Both coffee and tea contain caffeine, an alkaloid which serves as a stimulant. Cocoa contains theobromine, which is similar to caffeine.

Coffee (Rubiaceae)

2 Coffee is made from the ground seeds of the genus *Coffea*, which contains about 25 species. The commercial coffee industry is based primarily upon *C. arabica* (Arabian coffee), *C. canephora* (Robusta coffee), and *C. liberica* (Liberian coffee) with Arabian coffee accounting for about 90 percent of the world crop. *C. arabica* is a small tree from 4.5 to 9.0 m in height when unpruned but often held at 2 m by

Common species of coffee

Harvesting

pruning to facilitate harvest. The fruits are two-seeded drupes called cherries. The fleshy berries change from green through yellow to red. Harvesting of the mature berries has been by hand, but attempts to mechanize the harvest are under way. The two greenish-gray seeds are surrounded by several layers of tissue which must be removed by a combination of machine separation, fermentation, curing, milling, and polishing. The coffee beans must be roasted at a high temperature to develop the characteristic aroma, flavor, and color. Many commercial coffees are blends of different types and sources to provide a particular character. The final step is grinding the beans.

Many commercial coffees are blends

Growing conditions

3 Coffee production is limited to tropical and subtropical regions and is heavily concentrated in South and Central America and Africa; Brazil is the leading producer. Coffee is best adapted to an area with temperatures of 15 to 25°C, 200 to 300 cm of annual rainfall, and also a 2- to 3-month dry season, during which flowers are initiated. World coffee production is from 3 to 4 million metric tons. The only state in the United States growing coffee commercially is Hawaii, with annual crops of about 1000 metric tons.

Tea (Theaceae)

4 The most widely used caffeine-containing beverage is tea, made from the dried leaves of *Thea sinensis*. Allowed to grow to its full size, the tea plant is a small tree, but commercially it is kept pruned to a shrub about 1 m tall to make harvesting easy. Harvest involves plucking young tender shoots consisting of two to four leaves and the terminal bud. This has been a hand operation for centuries, but mechanical harvesters are being tested. The number of pickings per season varies from three or four in areas where growth stops in the winter to 25 to 30 in warmer areas.

Harvesting

Making green and black tea

5 The harvested leaves are wilted, rolled under pressure, and then further dried. At this stage of preparation, the leaves are completely dry but still green and can be used as *green tea*. For *black tea*, the major tea of world trade, the leaves are rolled once, allowed to ferment, and rolled repeatedly during fermentation, so that the green color is lost and the flavor changes, the result varying with the type of leaf used and the length of time fermentation is allowed to proceed. The grade of the final product depends on which part of the shoot is used, 'orange pekoe' being the smallest leaf, 'pekoe' the second leaf, etc.

6 Tea production is centered in India, China, Japan, and parts of Africa. Best growth is obtained in a tropical to subtropical climate at elevations of 1 km in the tropics to as low as sea level in cooler regions. Tea has been propagated from seed, but now cuttings and budding are being used to increase uniformity.

This passage was taken from BARDEN *et al.*, *Plant Science*, McGraw-Hill

Growing conditions

Cacao (Sterculiaceae)

7 The cacao plant (*Theobroma cacao*) is native to tropical America, where it is widely grown today. Parts of Africa also provide an excellent climate for cacao. Commercial production is limited to a zone of 20° north and south latitudes at low elevations with 150 to 200 cm of evenly distributed rainfall per year. Leading countries of production are Ghana, Nigeria, and Brazil. The cacao tree is quite small, usually 5 to 8 m high, and thrives best when partially shaded by larger trees. Flowering and fruiting occur continuously. The fruits are large football-shaped pods or capsules from 15 to 23 cm long and 7 to 10 cm in diameter. The fruits are harvested by hand when fully ripe, usually at weekly intervals, and split open (Fig. 21-21). The contents are removed and fermented to eliminate the pulp surrounding the seeds. After drying, the seeds or beans are dried and bagged for shipment.

Harvesting

8 The beans are cleaned, sorted, roasted, and ground into an oily paste called bitter chocolate. It may be hardened and sold, or converted into sweet chocolate by adding sugar and spices or into milk chocolate by adding milk as well. Cocoa is made by extracting most of the fat (cocoa butter), leaving a residue which is ground into powder.

G. Check your understanding

1. Look at paragraph 1 again.
- Which beverage does not contain caffeine?

2. Look at paragraph 2 again. Complete these statements correctly.
- 90% of the world coffee crop comes from
  - *Coffea arabica* ☐
  - *Coffea liberica* ☐
  - *Coffea canephora* ☐
- Mature coffee berries are
  - red ☐
  - green ☐
  - greenish-gray ☐
  - yellow ☐

3. Look at paragraph 3 again. Note down three climatic conditions which suit coffee well.

____

____

4. Look at paragraph 4 again. Which words have the same meaning as:
- picking by hand
- at the end

5. Look at paragraphs 5 and 6 again. Which words have the same meaning as:
- drooped
- again and again
- class of things of the same kind or quality
- heights

6. Look at paragraphs 7 and 8 again. Note down the best climatic conditions for growing cacao.

____

____

Can you explain the following words:
- pod
- pulp
- residue

H. Understanding discourse

Listen to this lecturer giving you some techniques to find out about before the next lecture. Note down the question and techniques you are to find out about.

FARM MACHINERY (1)

UNIT 9

A. Understanding a printed text (1)

The following text will introduce you to the topic of **farm machinery**. Look at the way it is divided into sections and paragraphs. Pay attention to the headings and notes in the margins, and to the illustrations and captions.

Now look at these questions:

1. What is the purpose of the first section?
2. Which types of tractor does the writer describe?
3. Which types of tractor are not illustrated?
4. What is the 'operating cycle'?
5. Which type of tractor is most commonly used in your country?

Read the passage through and find the answers to the questions. Remember, you do not have to understand every word to answer the questions.

The Agricultural Tractor

General

1 The importance of agricultural tractors in the world today should never be underestimated. They provide a great source of power which has almost entirely replaced the power of horses, and of man, for the many heavy and time-consuming tasks carried out on the land. *The importance of the tractor*

2 This power is being used to produce food for the nations of the world and there is no more important task than this. A tractor can do the work of numerous horses, and do this work in the same time, without having to be rested to recover from fatigue. Provided reasonable attention is given to its lubrication and it is kept supplied with fuels, it will work on indefinitely. The driver, of course, will require his rest, but it is not unusual for tractors to work around the clock when relief drivers are available.

3 As more and more food production is required to feed the ever-increasing world population, the numbers of tractors used will increase to meet the demand. Great areas of land will yet be brought into food production and the tractor will supply the necessary power for the work.

Tractor Types

4 The present-day tractor is a most useful machine, capable of supplying its power to numerous farm tasks. The most power absorbing of these tasks is usually the basic cultivation of land; therefore the tractor is designed to be able to do this task whilst at the same time being amply powered for the many other field tasks it is required to do, such as drilling seed, top dressing, spraying, haulage, etc. *The uses of a tractor*

5 During the years since the First World War, and in particular during, say, the last 20 years, a great amount of progress has been made in developing a machine capable of operating efficiently a very wide range of implements and machines. Many devices have been incorporated in the mechanism of the tractor for this purpose.

6 The type of tractor used on the land depends on the type of work to be done.

Track-laying Tractors (Fig. 1)

7 These tractors are sometimes referred to as "crawlers" and are usually tractors with a large horse-power and capable of doing very heavy work. On farms, they may be used for pulling a five- or six-furrow plough or for heavy cultivation. Yet, on the other hand, there is a very small track-laying tractor of no more than 6 h.p. (horse-power) which is used for work on market gardens.

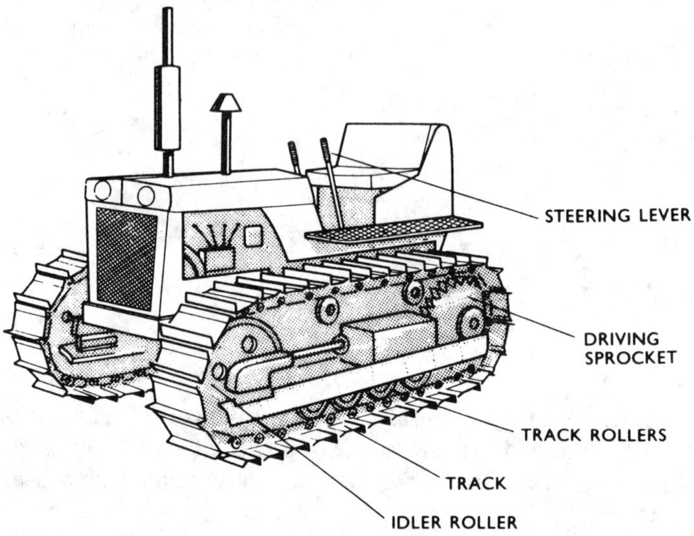

FIG. 1. A track-laying tractor.

Heavy Wheeled Tractors

8 Not only track-laying tractors are used for the heavy work on farms. Very large four-wheel-drive tractors are now common which are capable of pulling up to twelve furrow ploughs with engines of up to 500 h.p. These large tractors may be equipped with twin wheels all round, and no longer run in the furrow when ploughing, but run on top in the same way as a crawler tractor. They also carry out very heavy cultivations.

General-purpose Tractors (Fig. 2)

9 The general-purpose tractor is a wheeled tractor and the lighter type may be powered by an engine of about 40 h.p. whilst the other type may have an engine of 100 h.p. Either one or both of these types may be used on most farms and they are the most common type used today.

This text was taken from SHIPPEN *et al.*, *Basic Farm Machinery*,, Pergamon Press

▼

FIG 2. A general-purpose tractor.

Two-wheeled Tractors

10 Another type of tractor to be found on many small holdings and market gardens is the hand-operated. This type is powerful enough to do light cultivations and is steered by a walking operator.

11 Most tractors nowadays are powered by internal combustion engines, which, although they may vary slightly between one make and another as regards detail of construction, operate on the same basic principles.

The Internal Combustion Engine

12 The tractor's power is used for doing work in the field and also for driving stationary machines.

13 The power, which is produced by the engine, is transmitted through various mechanisms until it reaches the rear wheels which rotate. These mechanisms will be explained in later chapters. The power unit must be considered first.

14 In any internal combustion engine there are a number of essential parts which are so arranged that a basic series of events occur. This series of events is usually known as the operating cycle and most tractor engines operate on what is known as the four-stroke cycle.

15 The internal combustion engine is a form of heat engine and the name "heat engine" is given to it because heat energy, produced by the burning of fuel within the engine, is changed into mechanical energy. Different types of fuel are used in the different types of tractor engines but this does not alter the basic operation. The fuels used are petrol or propane for spark ignition, and diesel fuel is used in compression ignition engines. *The operating cycle*

B. Check your understanding

Now read the text carefully, looking up any new items in a dictionary or reference book. Then answer the following questions:

1. What sources of power have tractors largely replaced?
2. What does the tractor need to work on indefinitely?
3. Why will the number of tractors increase?

4. Which farm task uses most tractor power?

5. On what basis is the type of tractor chosen for use?

6. Where is the smallest track-laying tractor used?

7. What do heavy wheeled tractors and track-laying tractors have in common?

8. Which is the commonest type of tractor?

9. How are most tractors powered?

10. What does the internal combustion engine change?

C. Increase your vocabulary

In this section you should use your dictionary to help you answer the questions about the text.

1. Look at paragraphs 1–3 again. What words have the same meaning as:

- taking a lot of time
- completely
- without any time limit
- a great many
- thought too little of
- tiredness
- get better

2. Look at paragraphs 4–6 again. What words have the opposite meaning to:
- unable to
- not enough
- taken out of

3. Look at paragraph 7 again. Can you explain the words:

- heavy cultivation
- market garden

4. Look at paragraph 8 again. Can you explain the words:
- plough
- furrow

5. Look at paragraph 9 again. What word has the opposite meaning to:
- heavier

6. Look at paragraphs 10–11 again. Which words correspond to the definitions:
- general laws shown in the working of machines
- direct the course of a machine
- a small area of land usually farmed by the owner or tenant himself
- a little bit

7. Look at paragraphs 12–15 again. What words have the same meaning as:
- sent through
- change
- turn in a circle
- not moving

D. Check your grammar

DESCRIBING

1. Shorten the following sentences without changing their meaning, like this:

Weeding is a *task that consumes a great deal of time.*
Weeding is a *time-consuming task.*

- Farmers must feed a world population which is increasing all the time.
- You use a tractor which lays tracks in market gardens.
- Plants which produce fruits can be classified in many ways.

- The pineapple is a perennial which grows at a low level.
- Hawaii is an area that produces pineapples.
- The mango is a tree which bears fruits.
- The tsetse fly is an insect which carries disease.
- The banana is a herbaceous plant which grows fast.
- Some kinds of celery blanch themselves.
- Some plants are plants which pollinate themselves.

2. Shorten the following sentences without changing their meaning, like this:

Lettuce is a *crop which is harvested by hand.*
Lettuce is a *hand-harvested crop.*

- Snap bean is a crop which has large seeds.
- Celery is a crop which is propagated from seed.
- The pineapple has leaf bases which are shaped like funnels.
- The Muturu is a breed of cattle that has short horns.
- The papaya is a plant which has a single stem.
- Cucurbita is a squash which has a fine grain.
- Lettuce is an annual plant which has a shallow root.
- Some crops must be transported in cars which are cooled by ice.
- Most watermelons are of the type which has a red flesh.
- A spade is an example of a tool which is operated by hand.

E. Understanding a lecture

You are now going to hear part of a lecture. This time it is not divided into sections. The lecturer is going to tell you about the main parts of an internal combustion engine. You can see the parts below. The first time you listen, write down the letters that correspond to the different parts.

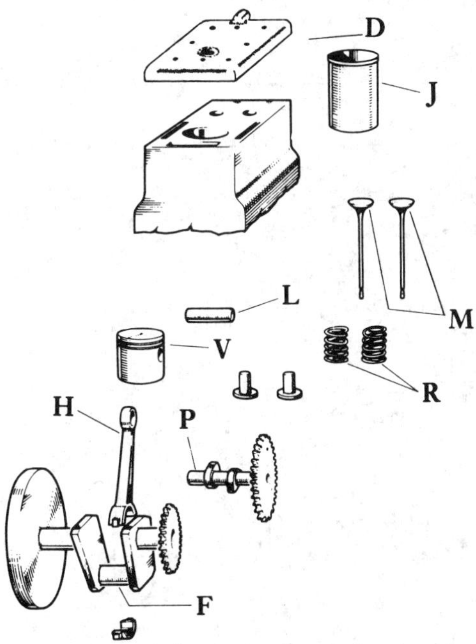

Section 1

- Write down the letters that correspond to these parts.

Cylinder block ____________

Cylinder head ____________

Cylinder liner ____________

Piston ____________

Crankshaft ____________

Connecting rod ____________

Gudgeon pin ____________

Valves ____________

Camshaft ____________

Valve springs ____________

Section 2

- Note down what the functions of the different parts are.

____

____

____

____

____

____

F. Understanding a printed text (2)

Read the following text carefully, looking up anything you do not understand.

TRACTOR ENGINES

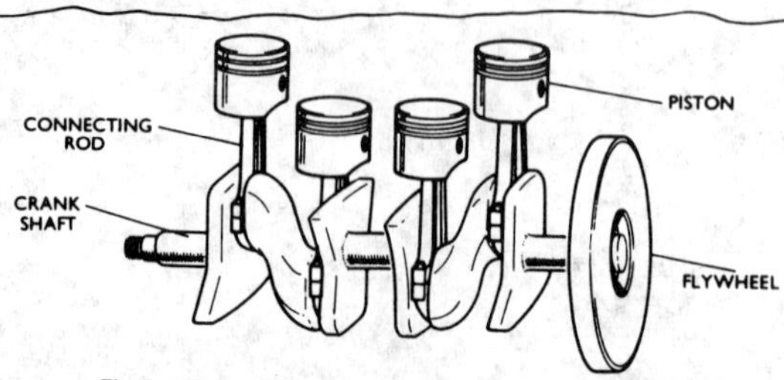

Fig. 6. A four-cylinder engine crankshaft with connecting rods and pistons.

1 Tractor engines are usually four-cylinder engines but it is not uncommon for them to have six cylinders; one, two and three cylinders are also used. Figure 6 shows a crankshaft with pistons and connecting rods for a four-cylinder engine.

2 Whether an engine is vertical, horizontal, or is single- or multi-cylinder, does not alter the method in which it operates. The basic cycle still takes place independently in each cylinder of the engine; which means that in a multi-cylinder engine, each cylinder with its piston, connecting rod and valves, etc., should be considered as an independent mechanical unit.

3 From Fig. 4 it will be seen that if the flywheel is turned, the crankshaft will also turn and this will cause the piston, which is free to move in the cylinder, to travel up and down. Because the gear wheel on the crankshaft is in mesh with the gear on the camshaft, the valves will also move up and down. If a thrust is applied to the top of the piston, the flywheel and crankshaft will again turn and in doing so will also operate the camshaft and valves.

4 This in fact is just what happens in the engine. Fuel/air mixture comes into the cylinder when the inlet valve is open, and when the valve closes, the mixture is compressed by the piston moving up. The mixture is then ignited and expands rapidly, and the expansion produces a downward thrust to the piston which in turn causes the crankshaft to rotate. The thrust is sufficient to keep the crankshaft turning so that the piston also continues to move up and down. The movement of these parts in relation to each other makes possible the operating cycle.

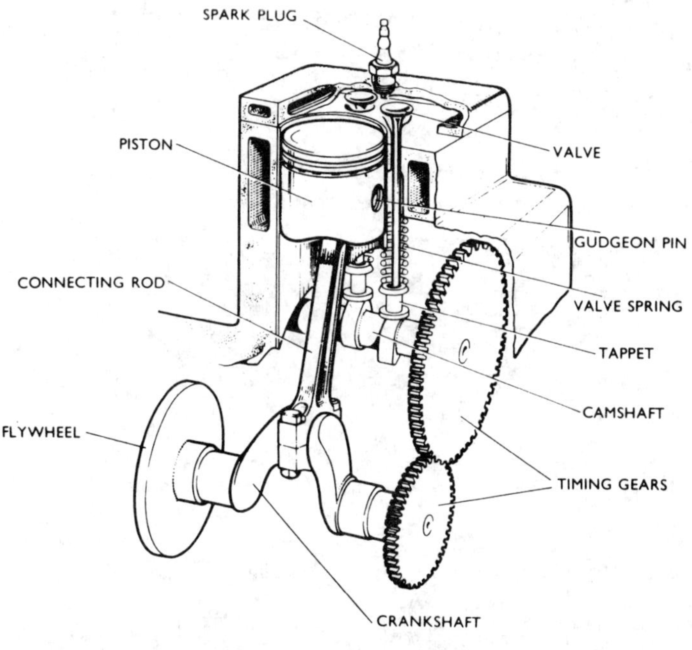

Fig. 4. A section through a single-cylinder engine showing the relative position of the working parts.

The Four-stroke Cycle (Spark Ignition)

5 Engines that are designed to run on petrol or propane require an electric spark to ignite the fuel/air mixture. These are known as Spark Ignition engines.

6 *No. 1. The induction stroke.* The piston moves down, inlet valve open, exhaust valve closed. This downward movement of the piston creates a partial vacuum inside the cylinder causing fuel/air mixture to rush in through the inlet valve passage. When the piston reaches the bottom of its stroke the inlet valve closes to prevent any escape of the fuel, therefore the fuel is trapped inside the cylinder.

7 *No. 2. The compression stroke.* The piston moves up, the inlet valve is closed, the exhaust valve is closed. This upward movement compresses the fuel/air mixture into a small space at the top of the cylinder. This space is usually provided by having the underside of the cylinder head suitably shaped, e.g. slightly concave.

8 *No. 3. The power stroke.* When the piston is at the top of the compression stroke with both valves closed, a spark is arranged to take place across the points of a spark plug which protrudes partly inside the combustion space. This spark immediately sets fire to the mixture which burns and expands, and in doing so forces down the piston; this provides the power.

9 *No. 4. The exhaust stroke.* The piston moves up the cylinder again, the inlet valve is closed, the exhaust valve is open. This upward movement pushes the burnt gases out through the exhaust valve outlet and to the atmosphere.

10 Figure 7 shows in diagram form the four strokes which make up the cycle of operation. The position of the valves as shown in the diagram is for the convenience of the reader. In practice they would be positioned side by side.

11 At the end of the exhaust stroke, the cycle starts again with a fresh charge of fuel/air mixture being taken into the cylinder.

12 Of the three strokes in the operating cycle, only one does the work. This is the third stroke which is the *power stroke.* The other three strokes make the power stroke possible, and the thrust of the power stroke keeps the crankshaft turning so that these strokes can take place. A stroke is one movement from top dead centre to bottom dead centre or vice versa.

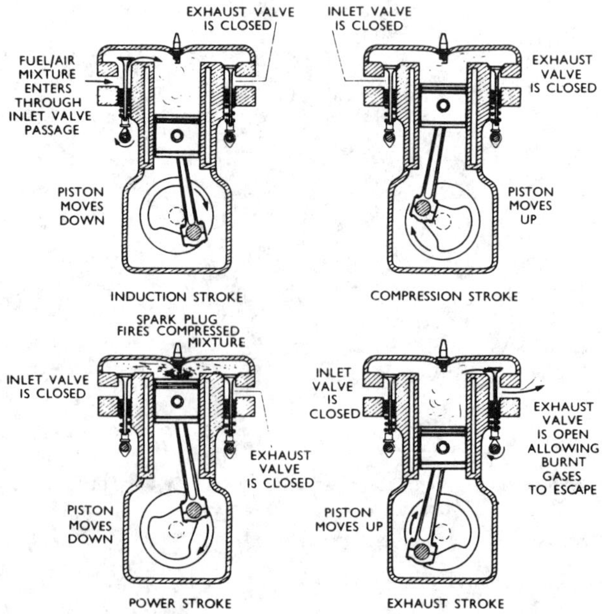

This text was taken from SHIPPEN *et al.*, *Basic Farm Machinery*, Pergamon Press

▼

G. Check your understanding

1. Look at paragraph 1 again. Are these statements correct or incorrect?
- Tractor engines never have six cylinders.
- Tractor engines can have 1, 2, 3, 4 or 6 cylinders.

2. Look at paragraph 2 again. Is this statement correct or incorrect?
- The method in which an engine works depends on the number of cylinders.

3. Look at paragraph 3 again. Is this statement correct or incorrect?
- Both the pistons and the valves move up and down.

4. Look at paragraph 4 again. Are these statements correct or incorrect?
- A fuel/air mixture enters the cylinder when the valves close.
- The piston moves up and down because the crankshaft keeps turning.

5. Look at paragraph 5 again. Is this statement correct or incorrect?
- Engines which use petrol or propane are called Spark Ignition engines.

6. Look at paragraph 6 again. Is this statement correct or incorrect?
- The engine part which stops fuel escaping is the piston.

7. Look at paragraph 7 again. Is this statement correct or incorrect?
- The fuel/air mixture is compressed into the piston.

8. Look at paragraph 8 again. Is this statement correct or incorrect?
- The mixture is ignited when both valves are closed.

9. Look at paragraph 9 again. Is this statement correct or incorrect?
- Burnt gases escape because of an upward movement.

10. Look at paragraphs 10–12 again. Is this statement correct or incorrect?
- The crankshaft is turned by all four strokes in the operating cycle.

H. Understanding discourse

A lecturer is going to give you three questions which you must work on. Note down the three questions.

____

____

____

____

____

____

UNIT 10 LIVESTOCK (2)

A. Understanding a printed text (1)

The following text will introduce you to the topic of **reproduction in farm animals**. Look at the way it is divided into sections and paragraphs. Pay attention to the headings and notes in the margins, and to the illustrations and captions.

Now look at these questions:

1. What is the text about?
2. In your own words, say what the oestrus cycle is.
3. What is the difference between Figure 15 and Figure 16?
4. In your own words, say what the second section is about.
5. Which animals does the writer refer to in the text?

Read the passage through and find the answers to the questions. Remember, you do not have to understand every word to answer the questions.

REPRODUCTION IN FARM ANIMALS

Topic 1. The Reproductive Mechanism

Hormones

1 Reproduction in farm stock is controlled by complex chemical substances called *hormones.* These are produced by various glands within the body and in addition to influencing sex and reproduction, some members of the hormone group influence growth and lactation.

Oestrus Cycle

Fertilisation

2 Farm animals start life when a female egg cell is fertilised by a male sperm. Although there are many thousands of undeveloped eggs on the female's ovaries (Figs. 15, 16), this fertilisation process can only take place at a definite period within the female's reproductive cycle, i.e. when one or more of these eggs has matured. It is, therefore, essential to mate her at the correct time.

Oestrus cycles

3 After a certain age, called puberty, the female's reproductive life is characterised by a series of what are called *oestrus cycles.* At the beginning of each of these oestrus cycles it is usual for one of the many thousands of undeveloped eggs to start to mature in the cow, and possibly two or even three in the sheep. . . . . .

Heat

4 When the eggs are mature the ovaries produce a hormone, *oestrogen,* which causes the female to come *on heat* or *in season*, and she will stand for service by the male. The eggs are shed into one of the fallopian tubes (Fig. 16) and pass on into a branch or horn of the uterus.

5 If successful mating has not occurred the eggs die and pass out of the reproductive tract unnoticed. A further oestrus cycle will then follow.

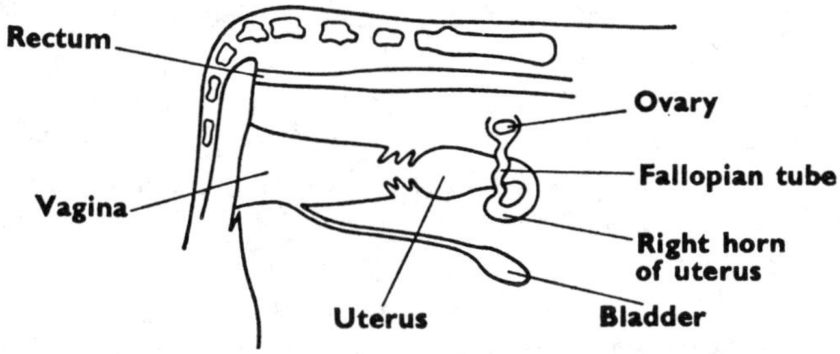

FIG. 15. Reproductive organs of the cow–side view.

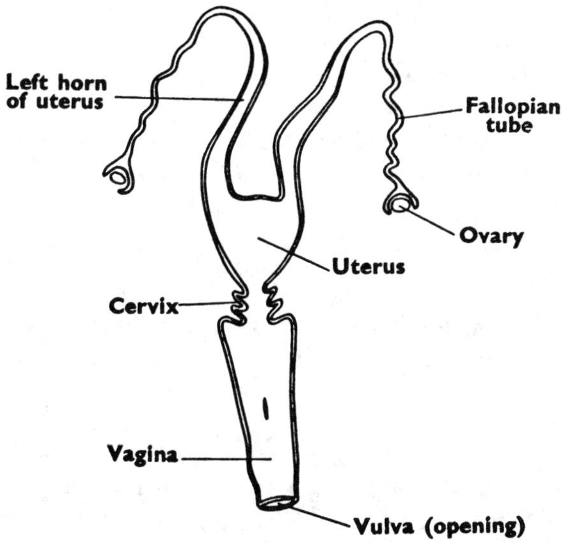

FIG. 16. Reproductive organs of the cow–top view.

Pregnancy

6 When mating does take place the male deposits many thousands of sperms (Fig. 17a) into the female vagina. These pass through the cervix, which protects the uterus or womb from external damage, and meet the eggs on their way down from the ovaries, but only one sperm is required to fertilise each egg.

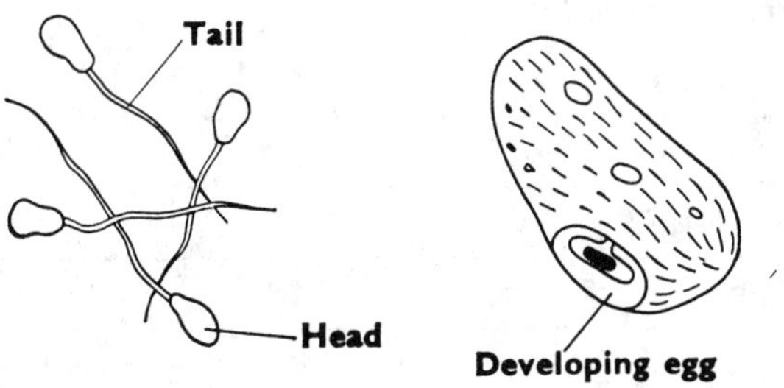

FIG. 17. (a) Sperms, (b) Ovary.

Successful fertilisation

7 When an egg is successfully fertilised, the cell which is produced divides many times and begins to grow into the young unborn animal. After a short period it becomes attached to the mother, within her uterus, by a navel cord. After this stage it becomes known as the embryo. Food substances can then pass from the mother to the offspring and waste products can travel in the opposite direction.

The demands of the unborn animal

8 During the initial stages of pregnancy the unborn animal makes little demands upon its mother. In the latter stages, during which it makes most growth, it requires a good supply of minerals for bone formation, and plenty of protein for the growth of internal organs and muscle.

9 The length of time between fertilisation and the birth of a young animal is known as the *gestation period.*

Variations between Farm Animals

Differences between animals

10 Although the basic principles of reproduction are the same within the different types of farm animals, there are one or two important differences in detail. In addition to the differences between the number of eggs shed at the mating period, the other main variations are as follows:

	Cow	Sheep
Usual length of heat	16–24 hr	27 hr
Average interval between heat periods	21 days	14–19 days
Gestation period	281–284 days or 9 months 1 week	144–150 days or 21 weeks

11 Heat periods occur at regular intervals throughout the year in healthy, non-pregnant, cattle.......... although they may be of relatively short duration in late winter. In British breeds of sheep, however, other than the Dorset Horn, oestrus cycles only occur in the autumn and early winter. This ensures that they lamb in the spring.

This text was taken from BUCKETT, *Introduction to Livestock Husbandry*, 2nd edn., Pergamon Press

B. Check your understanding

Now read the text carefully, looking up any new items in a dictionary or reference book. Then answer the following questions:

1. How are hormones produced?
2. When can fertilisation take place?
3. What happens at the beginning of the oestrus cycle?
4. Which hormone causes heat?
5. What happens after unsuccessful mating?
6. What prevents damage to the womb?
7. How is an unborn animal connected to the mother?

8. When does the unborn animal make most demands on the mother?

9. How many differences in reproduction are there between cattle and sheep?

10. Why does the oestrus cycle in British sheep usually happen in autumn or winter?

C. Increase your vocabulary

1. Look at paragraph 1 again. Which words correspond to these definitions?
- an organ in the body which separates substances from the blood
- act of producing milk
- having an effect on

2. Look at paragraphs 2 and 3 again. What words in the text could you replace with:
- during
- vital
- fixed
- develop fully

3. Look at paragraphs 4 to 6 again. What words have the same meaning as:
- without being seen
- dropped
- womb
- additional
- puts down

4. Look at paragraph 7 again. What words have the same meaning as:
- unwanted
- fixed to
- separates
- child; young

5. Look at paragraphs 8 and 9 again. Can you explain the words:
- protein
- muscle

6. Look at paragraphs 10 and 11 again. What words have the same meaning as:
- makes sure
- happen
- time between two or more events

D. Check your grammar

PARAGRAPH WRITING

Make complete sentences out of the following notes, putting the verbs in brackets in the right tense. Then arrange the sentences into two paragraphs. Make sure that the sentences follow each other logically and that each paragraph makes sense. Start with the heading:

INFERTILITY

- good/their/action/and/management/effects/prompt (help; minimise).
- egg/ovary/when/an/the/from (shed), cavity/the/left/a/by/in/the/egg/'yellow body' (grow)
- very/financial/infertility/farm/serious/the/to/stock/farmer/a/in/loss (represent)
- hormone/eggs/pregnant/in/this/more/animal/from/the (prevent; ripen)
- such/assistance/in/veterinary/cases (need)
- infertility/causes/of/several (know)
- it/animal/heat/again/sometimes/and/from/but/the/on (remain; prevent; come)
- function/yellow body's/a/the/hormone/main (be; produce)
- yellow body/if/the/normally/not/pregnant/the/animal (disappear; be)

E. Understanding a lecture

1. You are now going to hear part of a lecture, divided into short sections to help you understand it. As you listen, answer the questions below.

Section 1

- Is this statement correct or incorrect?
  We know all the causes of infertility in animals. ☐
- How many causes of infertility is the lecturer going to talk about?

Section 2

- Note down the following about the first cause of infertility:

  Name ____________________

  Reason for causing infertility ____________________

  ____

  Action to be taken ____________________

Section 3

- Note down the following about the second cause of infertility:

  Cause ____________________

  Part of animal affected ____________________

  Effect on embryo ____________________

  ____

  Other effects ____________________

  ____

Section 4

- Note down the following:

  Part of animal affected ____________________

  Visible symptom ____________________

  Action to be taken ____________________

  Effect of infection ____________________

Section 5

- Are these statements correct or incorrect?
- Protein deficiency is not a cause of infertility. ☐
- Iodine deficiency is a cause of infertility. ☐

Section 6
- What cause of infertility is the lecturer talking about?

____

Section 7
Are these statements correct or incorrect?
- A freemartin is always infertile. ☐
- The probable cause of a freemartin being infertile is the presence of male sex hormones. ☐

Section 8
Note down the possible effects of
- overworking your bull _______________

____

- resting your bull for several weeks _______________

____

____

2. Now turn your tape back to the beginning of the lecture and listen again. Take any further notes you want. When you have listened to the whole of the lecture, you will be asked to provide a short summary of it.

F. Understanding a printed text (2)

Read the following text carefully, looking up anything you do not understand.

The Mechanism of Inheritance

1 *Introduction.* Every farmer knows that if he crosses a Hereford bull with any other breed of cow, the white face of the Hereford will be passed to the offspring, but the rest of the calf's body may not necessarily be red like its father's. To understand this fact, and many others like it which have a large effect on the economics of animal husbandry, it is necessary to study something of the background of the mechanism by which factors such as colours are inherited.

2 *Chromosomes.* In each sperm and egg cell there is a small body, called a nucleus, which contains even smaller, rod-like objects, called chromosomes. Cattle and sheep produce different numbers of chromosomes,... ... but taking the example of cattle, the number within a bull's sperm and a cow's egg is the same. Also for each chromosome in the sperm there is a very similar one in the egg cell.

3 *Genes.* On each chromosome there are small areas which are called genes. These are responsible for controlling the different factors of inheritance. If it is assumed that the gene controlling hair colour is on chromosome A in the sperm, it will also be in the same place on the comparable chromosome A in the egg cell.

4 *Fertilisation and growth.* When a sperm fertilises an egg the embryo produced obtains the same number of chromosomes from the father as from the mother. For convenience we can call this number n. Thus the embryo will receive n chromosomes from the father and n chromosomes from the mother, making a total of $2n$ ($2n$ = number of chromosomes for that species, e.g. cattle $2n = 60$).

5 The single cell which is produced as a result of fertilisation has to divide many times before a new animal is produced. Each time a cell divides, the chromosomes inside its nucleus also divide, so that every new cell in the young animal's body contains the same number of chromosomes as the original fertilised cell. This is ordinary cell division (Fig. 18).

FIG. 18. Ordinary cell division.

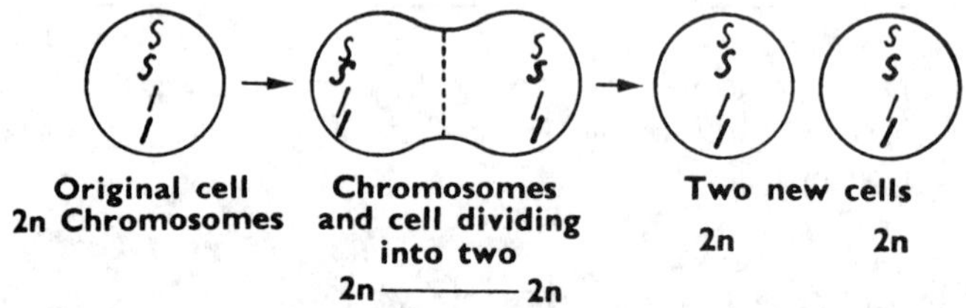

6 *Production of sperms and eggs.* When sperms or eggs are going to be formed the process is different. They are also formed by cell division, but of a different kind called reduction division. This type of division takes place in cells on the female's ovaries and in cells within the male's testes to ensure that the offspring has the same number, and not double the number, of chromosomes as each of its parents.

7 Taking the example of bull's sperms:

1. First the chromosomes in the reproductive cell come to lie together in matching pairs (Fig. 19(a)). (One chromosome of each pair will have been derived from the bull's father and the other from its mother.)

2. Then the cell and its nucleus divide, but instead of the chromosomes dividing as in ordinary cell division, one of the chromosomes from each of the pairs goes into one new cell and the other goes into the other (Fig. 19(b)). It is purely a matter of chance which of these chromosomes goes into which cell. This means that each new cell has n chromosomes and not $2n$.

FIG.19.
Production of sperms and eggs.

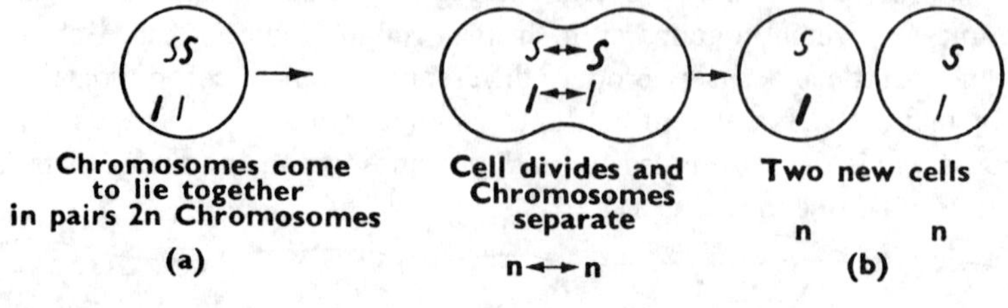

3. Actually a further division of the new cells takes place but this time the chromosomes themselves divide so that each sperm so formed still has *n* chromosomes. There are thus four cells from the original one.

8 The only difference between sperm and egg formation is that in the male each of these cells becomes a sperm, but nature has designed it so that only one of the four becomes an egg in the female.

9 *Examples.* Genes controlling a particular factor can be represented by letters. We can call the gene for red hair in cattle r, and for black hair B. Remember that after fertilisation two chromosomes are present which contain hair colour genes – one from the father, the other from the mother. From Fig. 19(b) it can be seen that when the animal is itself old enough to reproduce, its eggs can contain either of these two chromosomes. Therefore, we must represent each colour by two sets of genes, for example rr or BB.

10 The term homozygous is used to describe the situation where the two genes are the same, such as BB and heterozygous when they are different, such as Br.

EXAMPLE 1. Cross two pure Aberdeen Angus together

Sperms $B_1 / B_2 \times B_3 / B_4$ Eggs

Possible combinations: $B_1 B_3$ $B_1 B_4$ $B_2 B_3$ $B_2 B_4$

Result: All black.

EXAMPLE 2. Cross an Aberdeen Angus bull and a red cow

Sperms $B_1 / B_2 \times r_1 / r_2$ Eggs

Combinations: $B_1 r_1$ $B_1 r_2$ $B_2 r_1$ $B_2 r_2$

Result: Actually all are black, because the black genes are *dominant* to red genes, which are *recessive*, and the black colour overshadows red.

EXAMPLE 3. Cross two of these last offspring

Sperms $B_1 / r_1 \times B_2 / r_2$ Eggs

Combinations: $\underbrace{B_1 B_2 \quad B_1 r_2 \quad B_2 r_1}_{\text{3/4 Black}}$ $\underbrace{r_1 r_2}_{\text{1/4 Red}}$

Result: ¾ Black ¼ Red

11 Two important factors arise from this: (1) Some colours are dominant and overshadow others, for example the white face of a Hereford is dominant to other colours, black is dominant to red, and the genes for a polled factor are dominant to horns. Thus, if a Hereford is crossed with an Aberdeen Angus, the calves produced would be black, polled, but have a white head (Fig. 20). (2) There is an element of chance in reproduction. Thus, where there are many genes controlling a factor, such as milk yield, the production of the offspring cannot be accurately predicted.

The material in the grammar section, the lecture and the second text was based on BUCKETT, *Introduction to Livestock Husbandry*, 2nd edn., Pergamon Press

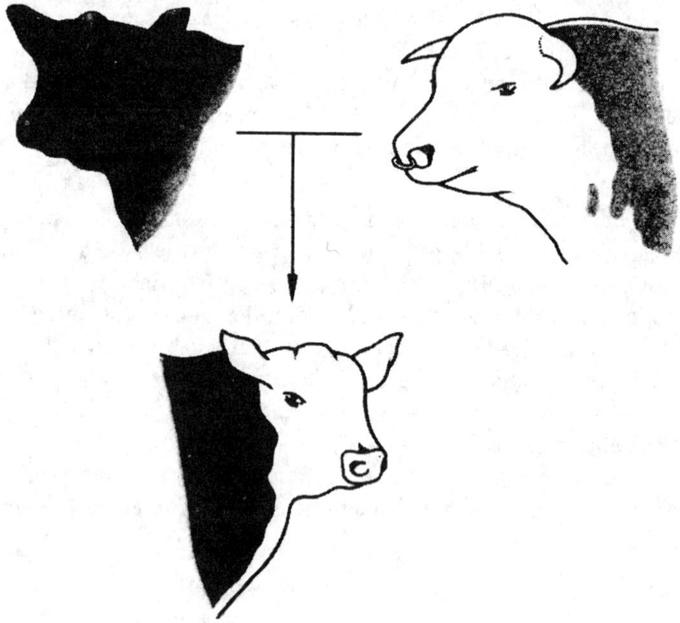

FIG. 20. Cross between Hereford and Aberdeen Angus.
Aberdeen Angus, black polled.
Hereford, horned, white face, red body.
Hereford Angus cross, polled,white face, black body.

G. Check your understanding

1. Look at the first paragraph again and say what these words refer to:
- line 1: he
- line 4: its
- line 5: it

2. Look at paragraph 2 again. Are these statements correct or incorrect?
- A chromosome is a small, rod-like object. ☐
- A bull's sperm and a cow's egg have different numbers of chromosomes. ☐

3. Look at paragraph 3 again. Which words correspond to the definitions:
- influences which cause something to develop
- something passed on from one's father and mother

4. Look at paragraph 4 again. What words have the same meaning as:
- so
- gets

5. Look at paragraph 5 again.
- Using your own words, explain Figure 18.

6. Look at paragraph 6 again. What word has the same meaning as:
- twice

7. Look at paragraph 7 again.
- Using your own words, explain Figure 19.
- Complete the statement correctly:
  In line 11, the second word *so* means
  therefore ☐
  in this way ☐
  with the result that ☐

8. Look at paragraph 8 again. Which words would you replace with
- with the result that

9. Look at paragraph 9 again. Can you explain why we represent each colour by two sets of genes?

10. Look at paragraph 10 again. Can you explain the words
- homozygous
- heterozygous

11. Look at paragraph 11 again. What words have the same meaning as:
- told in advance
- having control

H. Understanding discourse

Listen to this tutor giving some students instructions about examinations. Note down what the tutor says.

2

CHECK YOUR PROGRESS

The following short test is for you to check whether you are learning the skills you will need to study Agriculture. It is for *your* information. It is *not* an examination. You may use a dictionary.

A. Reading

Read the following text and then answer the questions.

Carrot (Umbelliferae)

The carrot (*Daucus carota*), an important vegetable throughout the world, is in the parsley family and is derived from wild forms native to Europe, Asia, and Africa. Its widespread use as food apparently dates back only to the sixteenth century. Before that, early forms of the carrot were used for medicinal purposes. It is thought that settlers introduced the carrot into Virginia in the early 1600s.

Of the total carrot crop, about 60 percent is grown for the fresh market, the remainder going to processing outlets. Carrots are also widely grown in home gardens and for local markets.

The plant is biennial; during the first year a rosette of feathery leaves and the storage root form. Temperature is particularly important for carrot production; a range of 15 to 21°C is optimum for both shape and color of roots. If the plant is left unharvested, a short stem bearing flowers is formed during the second growing season. The edible part of the carrot, a taproot, varies in length and diameter but ideally is 13 cm or more long and 2 to 4 cm in diameter. Many small secondary roots serve as absorbing organs.

Carrots are direct-seeded in the field. The small size and slow germination of seeds make it difficult to get an adequate stand without excellent seedbed preparation. The high labor cost of thinning makes precision seeding mandatory.

Carrots are harvested in many ways, but the trend is toward more and more mechanization. Most fresh-market carrots are washed, sized, hydrocooled, and packed in polyethylene bags. Although carrots can be stored for up to several months at 0°C and 90 to 95 percent relative humidity, long-term storage is hardly justified since they are harvested in certain areas almost year-round.

Beet (Chenopodiaceae)

The garden or table beet (*Beta vulgaris*), which belongs to the same species as the sugarbeet and Swiss chard, is believed to be native to Europe and North Africa. The plant as we know it is of quite recent origin.

This text was taken from BARDEN *et al.*, *Plant Science*, McGraw-Hill

Most beet production is for processing; total tonnage is about one-fifth that of carrots. Climatic requirements are similar to those for carrots and other root crops. Although beets will grow in warm weather, cool temperatures improve sugar content, internal color, and texture of the roots.

Like most other root crops, the beet is a biennial. However, after extended cool periods, some cultivars tend to bolt even before the roots reach a marketable size. The edible root is made up of alternating layers of vascular and storage tissues. Ideally, the color is uniform, but conducting tissues are usually lighter red than storage tissues.

The seeds, instead of being individual, are small fruits containing two to six seeds, which make precision seeding difficult. In home gardens, the seedlings can be thinned and used for beet greens, but thinning is not economically feasible on a commercial scale. Fortunately, small-sized roots are preferred for processing.

Answer these questions.

1. Complete the following statement correctly.

- The carrot belongs to the same family as
  - beet ☐
  - potato ☐
  - parsley ☐
  - *Daucus carota* ☐

2. Are these statements correct or incorrect?

- The carrot was not used at all before 1600. ☐
- Most of the American carrot crop is grown for processing. ☐

3. Complete the following statements correctly.

- If you do not harvest carrots in the same year that you sow them
  - the plants will die. ☐
  - the colour of the roots will change. ☐
  - the plants will flower ☐
  - the shape of the roots will change. ☐
- It is important to sow carrots precisely because
  - it costs a lot to thin the seedlings. ☐
  - the seeds are slow to germinate. ☐
  - the seeds are small. ☐
  - the seeds need a good seedbed. ☐

4. Is this statement correct or incorrect?

- It is not worth storing carrots for long periods of time.

5. Complete the following statement correctly.

- Swiss Chard
  - is exactly the same plant as beet. ☐
  - is also called *Beta vulgaris*. ☐
  - is a very recent plant. ☐
  - is in the same family as beet. ☐

6. Are these statements correct or incorrect?

- The best beet crops are grown in warm climates ☐
- Some beet cultivars bolt if the weather is too warm. ☐
- Thinning is not worthwhile for farmers who grow beet commercially. ☐

▼

B. Writing

Imagine you are an agricultural adviser. A farmer has asked you for advice about breeding cattle. Write to the farmer, advising him what to do. The text below will give you the information you need to write your recommendations.

Selection of Breeding Animals

Farmers endeavour to select the best animals to breed from, but the performance of an animal is not always a good guide to the performance of the offspring. There are two main reasons for this. The first is that an animal's performance is very much influenced by its environment, such as the food it receives, the diseases it contracts and the way it is managed. This is in addition to the influence of its genetic make up. The second reason is that many of the factors, or ***traits***, of economic importance are controlled by a large number of genes and not one gene. The permutations possible when genes are distributed during sperm and egg production are therefore enormous. Thus it is possible for the poorest characters of two excellent parents to come together to produce a poor offspring.

Some traits are very much more heritable than others and there is a greater chance that if such a trait is seen in the parents it will be seen in the offspring. A scale of 0–100% can be used to denote the heritability of traits although the terms high, medium and low heritability can be employed. Highly heritable factors include economy of liveweight gain and rate of liveweight gain; the medium category includes butterfat percentage from dairy cows and fleece weight in sheep; low factors include milk yield in cattle ... ... In general growth rates tend to be high, but fertility and ability to survive are low.

When selecting animals it is important to limit the number of traits being considered to those that are important. The larger the number of traits the greater the number of permutations possible and the more difficult it is to achieve progress. Quicker progress will be made with traits of high heritability.

It is obviously very important to select the correct sire since he can influence more offspring in a herd or flock than a single female. The proportion of the females selected for breeding replacements depends upon the number of replacements required. For example, with dairy cows the selection pressure can be higher in a herd with static numbers of cows, and where each cow has a long herd life, than in a herd which is expanding in size and where animals are culled after a limited number of lactations.

The time between when an animal is born and when it produces young itself is also important since it can influence the speed of improvement.

... ...

This text was taken from BUCKETT, *Introduction to Livestock Husbandry*, 2nd edn., Pergamon Press

Selection Techniques

Production records help to spotlight both the good and bad animals within a group. Stock showing superiority in those commercial qualities required can be used for breeding replacements.

Performance testing is the technique whereby measurements of the animal's own performance are used to give a guide to the possible performance of its offspring. It is of value for traits which are highly heritable and is used, in particular, to assess the value of potential boars and beef bulls. Special testing centres have been established where the young animals are taken and their growth rate, economy of food conversion into liveweight gain and other highly heritable factors are measured. The best animals are then selected for use as sires. They can later be progeny tested.

Progeny testing involves a careful study of the performance of an animal's offspring. Those sires whose progeny consistently perform better than their contemporaries produced by other sires are then selected for extensive use. Sires with poor progeny, are slaughtered. The technique is particularly valuable in testing for traits with poor heritability, such as milk yield, and for traits which can only be assessed after slaughter or measured in one sex only.

The main disadvantages of the system are that it is expensive and that it takes a considerable amount of time. Under one of the main progeny testing schemes operated by the Milk Marketing Board young bulls are sent to A.I. centres at about 12–15 months of age. Their semen is used with the aim of getting about 300 cows in calf to each bull. The young bulls are then "laid off" until they can be assessed from the milking ability and conformation of their daughters.

C. Listening

You will now hear part of a lecture. Listen to each section of the lecture twice. After you have listened to each section for the second time, answer the questions below. Here is the lecture:

Section 1

- The lecturer names five kinds of poultry. How many can you note down?

____

____

____

____

Section 2

Note down the information the lecturer gives you about the breed shown opposite. If the lecturer does not give you the information to answer the question, write *none*.

- Name of breed __________
- Place of origin __________
- Breeding habits __________
- Size of clutch __________
- Performance as mother __________
- Quality of meat __________

Section 3

Note down the information the lecturer gives you about the breed shown opposite. If the lecturer does not give you the information to answer the question, write *none*.

- Name of breed in picture ____________
- Type of chicken ____________
- Performance as layer ____________
- Performance as mother ____________

Section 4

Note down the information the lecturer gives you about the breed shown opposite. If the lecturer does not give you the information to answer the question, write *none*.

- Name of breed in picture ____________
- Type of chicken ____________
- Performance as layer ____________
- Performance as mother ____________
- Quality of meat ____________

Section 5

Note down the information the lecturer gives you about the breed shown opposite. If the lecturer does not give you the information to answer the question, write *none*.

- Name of breed in picture ____________
- Type of chicken ____________
- Performance as mother ____________
- Quality of meat ____________

This lecture was based on material from JOY & WIBBERLEY, *A Tropical Agriculture Handbook*, Cassell

A. Understanding a printed text (1)

The following text will introduce you to the topic of **forestry**. Look at the way it is divided into sections and paragraphs. Pay attention to the headings and notes in the margins, and to the illustrations and captions.

Now look at these questions:

1. What must forestry be based on?
2. Which two systems of classification does the writer mention?
3. Summarise what paragraph 3 is about in your own words.
4. Do all trees have the same kind of root system?
5. The writer describes one cause of damage to forests. What is it?

Read the passage through and find the answers to the questions. Remember, you do not have to understand every word to answer the questions.

Forestry

Principles of Forestry

Since the primary task of the forester is centered on producing trees, understanding trees and their growth is essential. Although certain special terms may be used for convenience, the principles underlying the growth, development, and reproduction of trees of the forest are the same as those for other crop species. Thus, forestry must be based on a sound understanding of botanical principles. One critical point to keep in mind is that trees are generally managed for wood production, i.e., stem tissue. Compared with most horticultural and agronomic crops, forests represent long-lived, perennial plants. (Orchards are somewhere between forests and most other crops.)

The basis of forestry

2 **Classification**

In United States forests, all trees are spermatophytes, or seed plants. They are classified formally as either gymnosperms (cone-bearing plants) or angiosperms (flowering plants). This formal c'assification is accepted scientifically, but others are routinely used by foresters. The most common general classification identifies a tree according to its leaf form as a hardwood or softwood tree. *Hardwoods* are broad-leaved, generally deciduous, flower-bearing trees (Fig. 25-2). *Softwoods* are needle-leaved, mostly evergreen, cone-bearing trees (Fig. 25-3). This broad classification is far from perfect: the southern long-needled pine is classified as a softwood species although its wood is harder than that of many of the hardwoods, and several deciduous broad-leaved species yield softwood, e.g., basswood, willow, and aspen. This leads to the confusing terms hard hardwoods and soft hardwoods. Some sense can be made from this when it is remembered that the terms were established early, when the demand was for the soft white pine and the very hard white oaks only. Later use of numerous species of widely varying hardness or softness could not erase the old established usage.

Gymnosperms and angiosperms

Hardwoods and softwoods

Morphology and Anatomy

3 Although a forest may contain annual, biennial, and short-lived perennial plant species, trees are generally considered to be the major vegetation; they are long-lived perennials that usually produce a single central stem and attain a height of 6 m or more. Rarely does a forest tree reach maturity in less than 15 to 20 years; some trees may grow continuously for centuries though generally very slowly after the first hundred years. In considering the growth of trees, increases in both height and diameter, or girth, are important in determining the yield of lumber or other products. In terms of gross morphology, as a rule, trees growing in a typical forest environment are taller and have smaller root systems than those growing separately or under widely spaced, ornamental settings. The apical or top portion of the tree, the *crown*, is also smaller under crowded conditions. Note that the word crown has a different meaning in forestry than in horticulture and agronomy. In forestry, the crown is that portion of the tree which has branches (Fig. 25-4). Since shade decreases the ability of branches to survive, the trees in crowded forests tend to shed their lower branches and thus have less crown in proportion to clear stem.

Growing time

Forest environment

The purpose of the root system

4 ***Roots*** Roots serve the same general function as for other plants: anchorage, absorption of water and essential minerals, and storage of photosynthate. The storage function is less important in forest trees than for most perennial field crops since stem tissue takes over much of the storage function in trees.

Differences in root systems

5 Tree species differ markedly in the extent of their root systems, differences that may affect how the species are managed in a forest. For example, most oak, hickory, and walnut trees have deep and extensive taproot systems which provide extremely solid anchorage. Thus, these and other trees with similar root systems are not as prone to uprooting by severe winds. In harvesting a forest, wind damage is a minor concern with these trees. Spruces and balsam, on the other hand, have shallow root systems and can be uprooted even by moderate winds. In planning the harvest, such trees must be removed or allowed to remain in a pattern that will ensure adequate wind protection. Most forest trees fall somewhere between these extremes. Generally, the root system provides sound anchorage except in relatively extreme conditions. Of course, even deep-rooted species can be uprooted under severe conditions, such as prolonged heavy rains before or during high winds. Hurricane conditions can level extensive areas of forest. In addition, trees that normally produce taproots may develop shallow root systems because of poor soil conditions, e.g., only a thin layer of soil above the bedrock. In such cases trees may be not only stunted but also easily blown over.

This text was taken from BARDEN *et al.*, *Plant Science*, McGraw-Hill

▼

B. Check your understanding

Now read the text carefully, looking up any new items in a dictionary or reference book. Then answer the following questions:

1. Why are forests usually managed?
2. What is the difference between an angiosperm and a gymnosperm?
3. How are hardwoods and softwoods differentiated?
4. What kind of tree is the aspen?
5. How long does it take for a forest to mature?
6. How do you assess the amount of wood you will get from a tree?
7. What is a crown?
8. Why is the storage function of a tree's roots less important than those of a field crop?
9. Why is an oak less likely to be uprooted than a balsam?
10. When is the root system inadequate to anchor a tree?

C. Increase your vocabulary

1. Look at the first paragraph again. What words have the same meaning as:
- well-informed
- main; most important

2. Look at paragraph 2 again. What words have the opposite meaning to:
- straightforward; easy to understand
- as an exception

3. Look at paragraph 2 again. Can you explain the words:
- hard hardwood
- hard softwood

4. Look at paragraph 3 again. What words have the same meaning as:
- go on living
- deciding
- hundreds of years
- part
- all the time
- seldom

5. Look at paragraphs 4 and 5 again. Which words correspond to these definitions:
- holding something down so that it is not blown over
- liable or inclined to
- going on for a long time
- poorly developed
- qualities which are as wide apart as possible

D. Check your grammar

1. ASKING QUESTIONS

You want to find out about the oil palm. Your teacher has the information you need. Find out about the oil palm by asking questions. Note down the answers. Ask about the following points:

Botanical name: ______________________

Family: ______________________

Varieties: ____________________

Seed rate: ____________________

Planting out: ____________________

Spacing: ____________________

Maturation: ____________________

Harvest: ____________________

Yields: ____________________

Composition percentage: ____________________

Diseases: ____________________

Processing: ____________________

2. PARAGRAPH WRITING

Now write a short paragraph on the oil palm, using the information you have noted down above.

E. Understanding a lecture

1. You are now going to hear part of a lecture, divided into short sections to help you understand it. As you listen, answer the questions below.

Section 1

- Is this statement correct or incorrect?
  The forest is an ecologically simple unit. ☐
- The lecturer mentions two kinds of vegetation that make up a forest. What are they?

- Apart from trees, what other factors should the manager consider?

Section 2

- Note down what a stand is.

- What term does the lecturer give a group of stands?

Section 3

- What factors in the site class affect the productivity of trees?

- Note down the two different kinds of stand.

Section 4

- Where is the diameter of a tree measured from?

- What measurements are used for the diameter of trees?

Sections 5 and 6

Complete the table below using the information the lecturer gives you. He does not give you all the information to complete the whole table.

Section 5

Section 6

Term	*Height*	*Diameter*

2. Now turn your tape back to the beginning of the lecture and listen to it again. This time, instead of answering questions, take notes. When you have listened to the whole of the lecture, you will be asked to provide a short summary of it. So make sure you note down the important points.

F. Understanding a printed text (2)

Read the following text carefully, looking up anything you do not understand.

Shade tolerance

Classification schemes

1 The most important of several broad ecological classifications of trees for forest management is shade tolerance. A *shade-tolerant tree* can generally withstand closer planting or more competition for light than a nontolerant type. This becomes a major consideration in planting, thinning, or harvesting operations and schedules. Forest trees are classified under three broad headings with regard to light requirements:

Shade-tolerant species—e.g., balsam fir, hemlock, redwood, basswood, spruce, birch, and maple

Intermediate—e.g., Douglas fir, ash, elm, and many oaks

Shade-intolerant—e.g., cypress, eastern red cedar, larch (tamarack), pine, aspen, black cherry, black walnut, cottonwood, hickory, locust, red gum, sycamore, yellow poplar, and willow

The common explanation for differences

2 Species in these three groups presumably differ in basic physiological processes related to photosynthesis. The most common explanation is that the differences in shade tolerance are related directly to differences in the amount of light required to reach light compensation, or the amount of light a plant needs for net photosynthesis to occur, as well as light saturation (see Chap. 7). It is important to note that these classifications are broad. Within any species are genotypes which may be more or less tolerant to shading than the average. Since a common cause of shading is crowding or dense planting, trees that tolerate shading are, within limits, also expected to tolerate more crowded conditions than nontolerant species. Tolerance is obviously a major consideration in deciding whether to manage trees as even-aged stands or as uneven-aged, stands where regeneration and growth must take place in the shade of several older age classes.

This text was taken from BARDEN *et al.*, *Plant Science*, McGraw-Hill

Crown development

3 The pattern of crown development is important in identifying the species that will succeed in uneven-aged conditions; related to, or an expression of, shade tolerance, it can be a significant factor in thinning, harvesting, and forest-management decisions. In a given stand, some trees, as they go through their life cycles, grow above all others and appear to dominate a stand. Others, at the opposite extreme, may never emerge at the upper surface of the forest crown or foliar canopy. Many fall somewhere in between, not truly dominant but not obviously occurring only at the lower levels of the canopy. Obviously trees that do not reach a dominant crown position must be shade-tolerant or else they would not survive and might as well be cut and used.

The importance of crown types

4 Recognizing crown types, shading, and spacing requirements are crucial in making harvest and thinning decisions. Four types of crown development and position are recognized:

Dominant Trees—The crowns extend above the general height of the forest canopy and receive full sunlight from above and partial sun from the sides. They experience minimal shading because they reach beyond competitors.

Codominant Trees—Crown development is at the general level of the forest canopy. Trees receive full sunlight from above but lateral shading from adjacent trees. Crowns are medium-sized and not as extensive as those associated with dominant trees.

Intermediate Trees—Crowns are formed below the general level of the forest canopy but extend up into it. Trees that develop in this manner must tolerate shading.

Overtopped Trees—Crowns are formed below the canopy level, receiving only filtered light.

Classification according to other ecological factors

5 Species can also be classified according to other ecological factors. Three species of pines, white, red, and jack, are adapted to the dry, sandy soils of the north. On heavier, well-drained soils in the same region, mixed hardwood forests are found, along with other pines, spruce, fir, and hemlock. Of course, in both situations, trees are adapted to the typical northern climatic conditions.

6 Other examples of widespread habitat are the extensive southern pine forests (see Fig. 25-7), typical of the southern coastal plains, the gums (Fig. 25-8) and cottonwoods found in association with southern bogs and swamps, and the cedars and balsam firs found in association with the colder, northern swamp areas.

G. Check your understanding

1. Look at paragraph 1 again. Are these statements correct or incorrect?
- A hemlock tree can stand more shade than an aspen.
- A tamarack needs more light than a redwood.

2. Look at paragraph 2 again. Which words correspond to these definitions:
- growing again
- the genetic makeup of an individual, determined by the assemblage of genes it possesses
- thick
- supposedly

3. Look at paragraph 3 again. Which words have the same meaning as:
- showing what something is
- having meaning or importance
- have control or influence over
- an overhanging covering

4. Look at paragraph 4 again. Can you explain in your own words the differences between dominant, codominant, intermediate and overtopped trees?

5. Look at paragraphs 5 and 6 again. Can you explain:
- habitat
- swamp

6. Choose one or more trees which are native to your country and with which you are familiar. Can you classify them according to the information given in the text? Write a paragraph or two classifying your chosen tree(s).

H. Understanding discourse

A tutor is telling some students about some arrangements he has made. Note down the arrangements that have been made and what you should do.

FARM MANAGEMENT (1)

UNIT 12

A. Understanding a printed text (1)

The following text will introduce you to the topic of **farm management**. Look at the way it is divided into sections and paragraphs. Pay attention to the headings and notes in the margins.

Now look at these questions:

1. What does the first section emphasise?
2. Where do differences in management show up?
3. What does the writer compare farm management with?
4. Do most definitions of farm management contain many of the same ideas?
5. Can you explain the writer's definition of farm management in your own words?

Read the passage through and find the answers to the questions. Remember, you do not have to understand every word to answer the questions.

Management

HIGHLIGHTS

1 Management is the key ingredient. The manager "makes" or "breaks" a business. Brains are more important than brawn. How often have you heard these or similar statements or read them in farm magazines? They all emphasize an important factor in the operation of a farm or ranch in today's world. *Management is important.* This does not mean that management was not important in the past. However, in an agriculture which is highly mechanized, uses many technological innovations, and operates with large amounts of borrowed capital, management takes on a new dimension and importance.

The importance of management

2 Why do some farmers make more money than others? Why do some farm businesses grow and expand while others stuggle to maintain their current size? Good or bad luck cannot explain all the differences observed in the profitability of farms and ranches even among those which have about the same amount of land and capital available. Farm business records from many states show the top one-third of the farms to be highly profitable while the bottom one-third are often operating at a loss. Why the difference? Observation and analysis often lead to the same conclusion. The difference is due to management.

Why some farms are different to others

3 These differences in management can show up in three areas: production, marketing, and financing. Production differences include the choice of agricultural commodities to be produced and how they are produced. Marketing includes the when, where, and how of purchasing inputs and selling commodities. Differences here are reflected in different prices paid and received. Financing covers not only borrowing money and the related questions of when, where, and how much, but also the entire area of how to acquire the resources necessary to produce agricultural commodities. There is risk to be considered in all three areas—how farm managers adapt to and handle this risk can have a major impact on profit.

Where the differences show up

4 If management is so important, we must ask even more questions. What exactly is management? What do managers do? What is the difference between management and labor? What knowledge and skills are needed to become a better manager? The answers to the first few questions are discussed in Chapter 1. The last question forms the basis for the remainder of this book.

Farm and nonfarm

The differences beween the two

Definitions of farm management

Common concepts

A definition

This text was taken from BUCKETT, *Introduction to Livestock Husbandry*, Pergamon Press

FARM AND RANCH MANAGEMENT

There are some differences in the management of a farm or ranch and the management of a nonfarm business corporation. The obvious differences are in size, type of business, and the products or services produced. Other differences include the relationship between labor and management and the setting of goals.

6 In a corporation,' the board of directors sets policies and goals and hires managers to achieve them. It is generally easy to separate corporation employees who are managers from those who provide the labor necessary to produce a product or provide a service. On a typical farm or ranch, one individual or a family group sets the goals, provides the management, and performs much or all of the labor required. This makes it difficult to separate the management activity from labor because the same individual(s) are involved and both tasks may be performed at the same time. For example, a farmer may be mentally planning a marketing strategy or developing next year's cropping program while operating a tractor or combine. When one person performs both the management and labor functions, there is a danger that the immediate need for labor to perform some task will place management in a secondary role, with management decisions postponed or delayed.

7 These differences are not readily apparent when definitions of business management and farm and ranch management are compared, but they are nevertheless important. There are many different textbook definitions of farm and ranch management, but several common points run through all of them. One of the more concise definitions is "Farm management is concerned with the decisions which affect the profitability of the farm business."[2] This broad definition contains several important points. First, it identifies profitability as a major objective of the business but not necessarily to the exclusion of other objectives. Second, this definition specifically identifies decisions and decision making as part of the management process.

8 Other farm management texts contain alternative definitions but with many of the same concepts or ideas. There is usually some reference to *decisions* or *decision making* as being part of the management process. Also, some mention is made of *goals* or *objectives*. This may be done in general terms, or a more specific goal such as profit maximization may be identified as being an important end to which the management effort is directed. Finally, some mention of the *organization* and *operation* of the farm or ranch business is included in many definitions.

9 Farm and ranch management can be thought of, then, as being a decision-making process. It is a continual process because of the continual changes taking place in our economy and in an individual business. The decisions are concerned with allocating the limited resources of land, labor, and capital among alternative and usually competing uses. This allocation process forces the manager to identify goals and objectives to guide and direct the decision making.

10 In this text, the following definition will be adopted: *Farm and ranch management is the decision-making process whereby limited resources are allocated to a number of production alternatives to organize and operate the business in such a way as to attain some objective(s).* While this definition is somewhat long, it does identify most of the characteristics of the management activities found on farms and ranches.

[2]Emery N. Castle, M. H. Becker, and F. J. Smith, *Farm Business Management*, 2d ed., The Macmillan Company, New York, 1972, p. 3.

B. Check your understanding

Now read the text carefully, looking up any new items in a dictionary or reference book. Then answer the following questions:

1. Why has farm management taken on a new dimension and importance?
2. Does luck account for the success or failure of farms?
3. What factor does account for the success or failure of farms?
4. In what three areas does the farm manager face risks?
5. What is different between a farm and a business corporation?
6. Why is it difficult to separate management and labour on a farm?
7. What danger does the farmer who is both manager and worker face?
8. What ideas do most definitions of farm management have in common?
9. How does the writer think of farm management?
10. Why does this process go on all the time?

C. Increase your vocabulary

1. Look at the first paragraph again. What words correspond to these definitions:

- a measurement of any sort
- something new that is introduced
- physical strength
- money with which a business is started or expanded
- something that is said

2. Look at paragraph 2 again. What words have the same meaning as:

- keep
- present, at the moment
- working

3. Look at paragraphs 3 to 4 again. What words have the opposite meaning to:

- minor
- exclude
- selling
- lending
- ignorance

4. Look at paragraphs 5 to 6 again. What words have the same meaning as:

- carries out, does
- reach, attain
- employs
- in the mind
- very clear, easy to understand
- put off until a later date

5. Look at paragraph 7 again. Can you explain the words:

- ranch
- profitability

6. Look at paragraph 8 again. What do you understand by:

- objectives
- the organisation of a farm
- the operation of a farm

7. Look at paragraphs 9 to 10 again. What words correspond to these definitions:

- giving as a share or for a purpose
- choice between two or more things
- by which

D. Check your grammar

Do you remember?

Some verbs must be followed by the -ing form:
He admitted *taking* my book.

Some verbs must be followed by the infinitive:
He agreed *to lend* me his book.

1. Make sentences from the following notes. Think whether you should use the infinitive or the -ing form after the verbs.

- He/suggest/read/chapter 5/before/we/go/lecture.
- He/want/spend/more time/study.
- The rain/not stop/fall/until yesterday evening.
- A good farmer/try/set/objectives.
- Peter/prevent/harvest/his crops/bad weather.
- I enjoy/listen/music.
- He/not mind/work/weekends.
- I/not want/risk/fail/my exams.
- He/forget/hand in/essay/last night.
- There/nothing/that shop/worth/buy.
- It/no use/leave/your work/last minute.
- I/miss/watch/college football match/last week.
- He/hope/get/good results/his examinations.
- I/expect/grow/more vegetables/next year.

2. Make nouns from the following adjectives, using *-ness* or *-ability*:

- complete
- useful
- profitable
- suitable
- comparable
- wet
- dry
- responsible
- hard
- divisible

E. Understanding a lecture

1. You are now going to hear part of a lecture, divided into short sections to help you understand it. As you listen, answer the questions below.

Section 1

Are these statements correct or incorrect?

- The lecturer does not expect the students to have read an introduction to farm management. ☐
- The students should know one definition of farm management. ☐
- Note down what kinds of activity the lecturer suggests farm management is. ☐

Section 2

- How many kinds of farm problems are there?

____

- Note down the first question.

____

Section 3

- Complete the following sentence with words from the lecture.

  The lecturer says production is decided by the _____ and _____ of inputs.

- Note down the first two inputs the lecturer mentions.

____

Section 4

- Note down the next question.

____

Are these statements correct or incorrect?

- Beef can only be produced in one way. ☐
- A good combination of inputs means a farmer can minimise his costs. ☐

Section 5

- Note down the next question.

____

- Is this statement correct or incorrect?
  There is only one reason why a farmer should choose a combination which suits him best. ☐

2. Now wind the tape back to the beginning of the lecture and listen to it again. This time, instead of answering questions, take notes. When you have listened to the whole of the lecture, you will be asked to provide a summary of it. So make sure you note down the important points.

F. Understanding a printed text (2)

Read the following text carefully, looking up anything you do not understand.

THE DECISION-MAKING PROCESS

1 The allocation of limited resources among a number of alternative uses requires a manager to make decisions. This is one of the reasons, if not the most important one, for some mention of decisions or decision making in a definition of farm and ranch management. Without decisions nothing would happen. Even allowing things to continue as they are implies a decision, perhaps not a good decision but a passive decision nevertheless.

How to make a decision

▼

2 The process of making a decision can be formalized into a logical and orderly series of steps. Important steps in the decision-making process are

1 Identify and define the problem.
2 Collect relevant data, facts, and information.
3 Identify and analyze alternative solutions.
4 Make the decision—select the best alternative.
5 Implement the decision.
6 Evaluate the results and bear responsibility for the outcome.

Following these steps will not ensure a perfect decision. It will, however, ensure that the decision is made in a logical and organized manner.

Identifying and Defining the Problem

Basic problems of the manager

3 Many problems confront a farm or ranch manager. Earlier in this chapter some problems were identifed in terms of deciding how much to produce, how to produce, and what to produce. These are basic problems faced by all managers. Problems also result from identifying something that is not as it should be. This may be a goal that is not being attained or a deficiency in the organization or operation of the business identified by finding a difference between *what is* and *what should be*. For example, a farmer may find that his cotton yield is 100 pounds per acre lower than the average for other farmers in the same county on the same soil type. This difference between what is, the farm yield, and what it should be, at least the county average yield, identifies a problem that needs attention.

Problems represent an opportunity

4 A manager must constantly be on the alert to identify problems and to identify them as quickly as possible. Most problems will not go away by themselves and represent an opportunity to increase the profitability of the business through wise decision making. Once identified, the problem should be concisely defined. Good problem definition will minimize the time required to complete the remainder of the decision-making steps.

Collecting Data and Information

Where to get data

5 Once a problem has been identified and properly defined, the next step should be to gather data, information, and facts and to make observations which pertain to the specific problem. A concise definition of the problem will help identify the type of data needed and prevent time being wasted gathering information which is not useful to the particular problem. Data may be obtained from a number of sources, including the local county extension office; bulletins and pamphlets from state experiment stations and agricultural colleges; dealers; salespersons of agricultural inputs; radio; TV; farm magazines; and neighbors. An important source of data and information is an accurate and complete set of past records kept for an individual business. When available this is often the best source. Whatever the source, the relative accuracy and reliability of the information obtained should be considered.

6 Decisions typically require information about future events since producing crops and livestock takes time. Information about future prices and yields is often needed and may require the decision maker to formulate some "estimates" or "expectations." Past prices and yields provide a starting point but will often need to be adjusted based on the best current information and the judgment of the decision maker.

It is important to make a distinction between data and information. Data can be thought of as an unorganized collection of facts and numbers obtained from various sources. In this form, the data may have very little use. To be useful, these data and facts need to be organized, sorted, and analyzed. Information can be thought of as the final product obtained from analyzing data in such a way that useful conclusions and results are obtained. Not all data need to be organized and summarized into useful information, but most are not useful until some type of analysis is made.

Data and information are not the same

8 Gathering data and facts and transforming them into information can be a never-ending task. A manager may never be satisfied with the accuracy and reliability of the data and the resulting information. However, this step must be terminated at some point to make it possible to move to the third step. It is important to remember that gathering data has a cost in terms of both time and money. Too much time spent gathering and analyzing data may result in a cost which cannot be justified by the extra income received from continual refinement of data collection and information processing.

Identifying and Analyzing Alternatives

List the alternatives

9 Once the relevant information is available, the manager can begin listing alternatives which are potential solutions to the problem. Several may become apparent during the process of collecting data and transforming data into information. Others may take considerable time and thought, and all possible alternatives should be considered. This is the time to brainstorm and list any idea which comes to mind. Custom, tradition, or habit should not restrict the number or types of alternatives which are considered.

Analyse the alternatives

10 Each alternative should be analyzed in a logical and organized manner to ensure accuracy and to prevent something from being overlooked. The principles and procedures discussed in the next five chapters provide the basis for sound analytical methods. In some cases, additional information may be needed to complete the analysis. Good judgment and practical experience may have to substitute for information which is unavailable or available only at an additional cost that is greater than the additional return from its use.

Making the Decision

It is not easy to choose the best solution

11 Choosing the best solution to a problem is not always easy, nor is the best solution always obvious. Sometimes the best solution is to do nothing or to go back, redefine the problem, and go through the decision-making steps again. These are legitimate decisions, but they should not be used as a way to avoid making a decision when a promising alternative is available.

Complications

12 After all the pros and cons of each alternative are weighed, one may not appear to be definitely better than any other. The one showing the greatest increase in expected profit would normally be selected. However, this selection is often complicated by uncertainty about the future, particularly future prices. If several alternatives have nearly the same potential effect on profit, the manager must then assess the chances or probability that each will have the expected or identified outcome and the risk associated with it.

This text was taken from BUCKETT, Introduction to Livestock Husbandry,
2nd edn.,Pergamon Press

Be careful of doing nothing

Implementing the Decision

13 Selecting the best alternative will not give the desired results unless the decision is correctly and promptly implemented. Resources may need to be acquired and organized, which requires some physical action to be taken. Of course, doing nothing is an alternative and a potential solution to a problem. However, the manager who decides to do nothing or to implement no new alternative should do this only after enough analysis of the problem to be sure that this is the correct decision. Too often doing nothing results from not allocating enough time to the decision-making process or a reluctance to make a decision.

Managers must bear responsibility

Evaluating Decisions and Bearing Responsibility

14 Responsibility for the outcome of a decision rests with the decision maker. A reluctance to bear responsibility may explain why some individuals find it so difficult to make a decision. However, since it is difficult for managers to avoid decision making, it follows that they must bear the responsibility. It goes along with the job.

Learn from your mistakes

15 Not every decision will be a perfect one. For this reason the results of each decision should be carefully evaluated to gather any data or information which can be used to modify the decision and to assist in making better decisions. Careful observation and analysis will result in additional information, allow corrections to be made, and improve future decisions. Managers who learn from their past mistakes may soon find people talking about their "good luck." While good luck may occasionally bless some managers, most is the result of hard work and good decisions.

G. Check your understanding

1. Look at paragraphs 1–2 again. Which words correspond to these definitions:
- connected with what is being thought about
- decide the amount or value of something
- choose
- carry out, put into practice
- suggests without actually saying

2. Look at paragraphs 3–4 again. What words have the same meaning as:
- be face to face with
- reached, achieved
- all the time
- briefly and to the point

3. Look at the sources of data available to the farmer in the USA in paragraph 5.
- Say which of these sources are available in your country.
- Are there other sources available in your country?

4. Look at paragraph 6 again. Is this statement correct or incorrect?
- A manager's own opinion plays no part in making an estimate.

5. Look at paragraph 7 again. Explain in your own words the difference between data and information.

6. Look at paragraph 8 again. Is this statement correct or incorrect?

- You should go on transforming data into information until you are satisfied it is completely accurate and reliable.

7. Look at paragraph 9 again. Is this statement correct or incorrect?

- Custom should not decide how many alternatives are considered.

8. Look at paragraph 10 again. What words have the same meaning as:

- sensible, reliable
- take the place of

9. Look at paragraphs 11–12 again. Are these statements correct or incorrect?

- Doing nothing can sometimes be the best decision to make.
- Doubt about future prices makes choosing the best alternative difficult.

10. Look at paragraph 13 again. What words have the same meaning as:

- obtained
- unwillingness

11. Look at paragraphs 14–15 again. What words have the opposite meaning to:

- make worse
- curse

H. Understanding discourse

Listen to this lecturer making an announcement. Note down what you yourself must do.

UNIT 13

VEGETABLES (2)

A. Understanding a printed text (1)

The following text will introduce you to the topic of **vine crops**. Look at the way it is divided into sections and paragraphs. Pay attention to the headings and notes in the margins.

Now look at these questions:

1. Are most vine crops cultivated in the same way?
2. Does the writer say where cucumbers came from?
3. Does the writer say what size cucumbers should be for the market?
4. Does the writer indicate the temperature at which muskmelons should be grown?
5. Are there any differences between cucumber and muskmelon production in your country and the USA?

Read the passage through and find the answers to those questions. Remember, you do not have to understand every word to answer the questions.

VINE CROPS (CURCUBITACEAE)

1 The vine crops include cucumber, muskmelon, watermelon, pumpkin, and squash. Each is a tender annual crop grown for its fruit. Cultural requirements for all are similar.

Cucumber

Origins

2 The cucumber (*Cucumis sativus*) is a popular vegetable cultivated since very early times, presumably well over 3000 years. Its origin was probably in India; it was known by the Romans and was brought to the New World by Columbus.

Temperature for germination

3 Cucumber, like squash, thrives under somewhat cooler temperatures than muskmelons or watermelons, but germination will not occur unless temperatures are at least 13°C and germination is much faster at 30°C. Since the cucumber reaches harvest maturity in a relatively short growing season and will grow at relatively cool temperatures, it can be grown in most of the United States. However, for success in the far north it may need to be transplanted.

Greenhouse production

In Europe, greenhouse cucumber production has been a large industry, especially in the northern areas. The English and Dutch developed cultivars which are long (often 0.6 m) and parthenocarpic for this type of production.

The plant

4 The plant is an annual trailing vine with a spiny or hairy stem. Cucumbers are monoecious, but recently introduced hybrids are gynoecious, producing female flowers only. For commercial plantings, a small percentage of a monoecious type is included to provide pollen. Gynoecious cultivars are used because of earliness, concentrated ripening, increased yields, and suitability for mechanical harvesting.

5 Most cucumbers fit into one of two categories. Slicing cultivars grown for the fresh market should be 12 to 20 cm long with a diameter of 4 to 5 cm and dark green. To supply top-quality cucumbers, the fields are picked every 2 to 4 days. The fruit of the pickling cultivars are smaller, lighter in color, and more numerous (Fig. 20-10). *The two categories*

6 Cucumbers are usually direct-seeded in the field after the danger of freezing weather is past. For once-over harvest, high plant populations of gynoecious types are used. With the multiple harvests normal for the fresh market, plant populations are lower. *Sowing*

Muskmelon

7 The muskmelon (*Cucumis melo* var. *reticulatus*), or cantaloupe, is thought to have originated in Asia, whence it eventually spread throughout Europe and the New World. The muskmelon is distinct from some other melons for its netted skin and its much better keeping quality. Another member of this group, the Persian melon, also has netted skin but is larger. The muskmelon fruit is 11 to 14 cm in diameter and has thick orange flesh and numerous seeds in the internal cavity. *Origins*

FIGURE 20-10
Modern cucumber cultivars are gynoecious and thus produce heavy, concentrated yields. This is a pickling-type cultivar. (*Harris Seed Company.*)

8 The plant is a tender, trailing vine which needs warm temperatures and plenty of sunshine. Low humidity and low rainfall make it easier to control fungal diseases, which are particularly severe on muskmelons in humid climates. Improved disease resistance is being incorporated into new cultivars. Most of our crop comes from irrigated regions in the southwest, where furrow irrigation is commonly used to avoid wetting the foliage. *The plant* *Irrigation*

9 For large plantings, muskmelons are direct-seeded but for small areas or where the growing season is too short, seed may be started in greenhouses and transplanted. In many areas muskmelons are grown on black plastic mulch, to which they are particularly responsive. *Sowing*

10 Maturity is critical if top-quality muskmelons are to be offered to the consumer. Because a muskmelon does not increase in sugar level after harvest, immature fruits are not only hard but insipid as well. *Maturity*

The most widely used index of maturity is how easily the fruit can be separated from the vine. A full-slip melon breaks from the vine cleanly, leaving no stem attached (Fig. 20-11). This provides a melon of excellent quality if handled properly, shipped under refrigeration, and marketed promptly. Half-slip melons have some stem remaining on the fruit and are of inferior quality even after softening.

Harvesting

11 Although many machines to harvest muskmelons have been tested, most fruit is still picked by hand. Muskmelons should be hydrocooled or iced down as soon as possible and held at 4 to 7°C during shipment. They should not be stored for more than a few days.

Other varieties

12 Other melons closely related to muskmelon but classified as *C. melo* var. *inodorous* include the honeydew, a somewhat larger, smooth-skinned melon with a very pleasant, sweet, greenish-white flesh, and Crenshaw and Casaba; some of these are shipped but many are consumed locally in the west.

FIGURE 20-11
When a muskmelon is mature (the full-slip stage), the stem will separate completely from the fruit and fruit quality is excellent.

This text was taken from BARDEN *et al.*, *Plant Science*, McGraw-Hill

B. Check your understanding

Now read the text carefully, looking up any new items in a dictionary or reference book. Then answer the following questions:

1. Does cucumber grow well at the same temperature as the muskmelon?
2. Who developed greenhouse cultivars for greenhouse production?
3. How does the writer define the cucumber plant?
4. How does the writer describe the two main categories of cucumber?
5. When are cucumbers direct-seeded in the field?
6. Where did muskmelons originally come from?
7. How is the muskmelon different from other melons?
8. Why is furrow irrigation used for muskmelon production?
9. How do you know that a muskmelon is mature?
10. How are most muskmelon harvested?

C. Increase your vocabulary

1. Look at the first paragraph again. What words have the same meaning as:
- not strong
- almost the same

2. Look at paragraph 2 again. What word has the opposite meaning to:
- not liked by most people

3. Look at paragraph 3 again. Can you explain the words:
- germination
- transplanted
- greenhouse
- parthenocarpic

4. Look at paragraph 4 again. What words correspond to the definitions:
- type of sex expression in which female flowers only are produced
- type of sex expression in which male and female structures are produced on the same plant
- growing along the ground
- a fine powder grown on flowers which fertilises other flowers

5. Look at paragraphs 5 and 6 again. Can you explain:
- the difference between a once-over harvest and a multiple harvest

6. Look at paragraph 7 again. What words have the same meaning as:
- a hollow space
- can be distinguished
- from where

7. Look at paragraph 8 again. What words have the opposite meaning to:
- dryness
- rarely
- upright
- taken out of

8. Look at paragraph 9 again. Can you explain the words:
- transplanted
- mulch

9. Look at paragraph 10 again. What words have the opposite meaning to:
- tasty
- broken off
- not important
- superior

10. Look at paragraphs 11 and 12 again. Can you explain the words:
- hydrocooled
- greenish-white

D. Check your grammar

Make complete sentences out of the following notes, putting the verbs in brackets in the right tense. Then arrange the sentences into three paragraphs. Make sure that each paragraph makes sense, and that the sentences follow each other logically. Start with the heading:

WATERMELONS

- Watermelons/worldwide/today (grow)

___

- Plant/cut/tender/deeply/with/leaves/a/vine/the (be)

___

- Also/but/the/Mississippi Valley/it/in (grow), first/the/Europeans/when/ (arrive)

___

- Fruits/to/in/round/the/oblong/4/and/25g/over/from/weight/to (be; range)

▼

- Usually/in/Africa/the/watermelon (consider; originate)

- When/mature/watermelons/they/overripe/but/not (harvest; be)

- Warm/for/long/a/growing/watermelons/season (require)

- Hand/trucks/and/or/they/by/trailers/on (harvest; load)

- Troublesome/but/humidity/detrimental/diseases/since/leaf/muskmelon/on/less/high/less/ than (be; be)

- Seeded types/recently/now/cultivars/only/seedless/the/market/on/until/but (be; introduce)

- Monoecious/it/bees/usually/by/and (be; pollinate)

E. Understanding a lecture

1. You are now going to hear part of a lecture, divided into short sections to help you understand it. As you listen, answer the questions below.

Note down the following.

Section 1

- Botanical name
- Family name
- Local name

Section 2

- Varieties
- Characteristics

Section 3

- Propagation
- Spacing
- Germination

Section 4

- Maturity
- Harvest

Section 5

- Pest
- Action to be taken
- Pest
- Action to be taken
- Pest
- Action to be taken

2. Now wind the tape back to the beginning of the lecture and listen to it again. This time, instead of answering questions, take notes. When you have listened to the whole of the lecture, you will be asked to provide a summary of it. So make sure you note down the important points.

F. Understanding a printed text (2)

Read the following text carefully, looking up anything you do not understand.

BEANS AND PEAS (FABACEAE)

1 The two major types of bean used as vegetables are the snap bean (*Phaseolus vulgaris*) and the lima bean (*P. lunatus*), both legumes. *P. vulgaris* is native to North and South America and was taken to Europe by early explorers.

The two major types

Snap Bean

2 This group includes snap beans and beans used in a mature stage (dried beans; see Chap. 19). Some cultivars can be used in the immature stage as a snap bean or allowed to mature and used as a shell bean. Although most snap beans are green, there are also yellow cultivars called *wax beans.*

The importance of snap beans

3 Snap bean ranks among the top 10 or 11 vegetables. More than 80 percent of the commercial crop is processed; the remainder is marketed fresh. Climatic conditions best for snap beans are mild temperatures and adequate but not excessive moisture. Since they require a relatively short growing season, several crops can be grown to provide continuous harvest in areas where temperatures do not become excessive.

4 The plant, a tender annual, is planted after danger of freezing has past and soil temperatures have reached at least 16°C to ensure germination. Beans are characterized as either bush or pole types. Bush beans grow on a low, erect plant which does not need support and seldom is taller than 50 cm. Pole beans grow as a vine, which must be supported as they grow to a length of about 2 m. Because of higher yields and distinctive flavor, pole beans are widely grown in home gardens.

The plant

5 Beans are direct-seeded in the field. From seeding to harvest takes 50 to 60 days for bush beans and 60 to 70 days for pole beans. Snap bean harvest is completely mechanized for bush types (Fig. 20-13). Pole beans are productive over a longer time but have declined markedly in commerce since they require hand harvest.

Seeding

6 Maturity of the beans at harvest is critical for consumer satisfaction. Optimum quality is obtained when the pods are almost fully grown but the beans are only about 25 percent of mature size. Pods allowed to mature further tend to develop toughness and fibers or strings, a characteristic plant breeders have essentially eliminated if beans are harvested at the proper stage. For the fresh market, snap beans are hydrocooled and shipped in refrigerated trucks.

Harvesting

▼

Lima Beans

The two types

7 Lima beans are of two distinct types, the small-seeded baby lima and the large-seeded standard lima. Although there is not complete agreement, most people include both types in *P. lunatus*. Each type has both vine and bush forms.

The plant

8 The plant resembles that of the snap bean in growth habit. Lima beans require a longer growing season than snap beans because they are grown for the seed rather than the pods. Nearly all commercially grown limas are of the bush type. All processing limas are harvested by machines, which mow off the vines and shell the beans mechanically so that only the seeds are hauled from the field.

Peas

Place of origin

9 The pea (*Pisum sativum*) is also called the garden pea or the English pea. Peas apparently originated in Europe or were at least well known to ancient Greeks and Romans; peas are widely grown in Europe today. Most cultivars in the United States are grown strictly for the seeds after shelling, but there are also many cultivars of sugar (edible-podded) peas.

Commercial peas

10 Almost all commercial peas are processed. Production is centered in the north because a cool growing season is required and as maturity is reached, the cooler the temperature the more leeway available for harvesting with optimum quality.

11 Commercial pea cultivars are of the bush type; pole or climbing types are confined to home garden use (Fig. 4-9). The pea is one of the hardy vegetables and is direct-seeded 2 to 3 weeks before the average last freeze. Early planting is desirable to ensure maturity before the heat of midsummer, which reduces both yield and quality.

Harvesting

12 Peas for processing are all mechanically harvested by mowing and are shelled by machine, requiring minimal labor. For the fresh market, peas are picked by hand, a slow, expensive operation which no doubt contributes to the scarcity of fresh peas on the supermarket shelf. Pea quality depends on both sugar content and texture. For maximum tonnage per hectare, one would have to wait well beyond the stage of top quality. Since peas should be young, green, tender, and sweet, the grower must balance yield and quality for maximum returns.

This text was taken from BARDEN *et al.*, *Plant Science*, McGraw-Hill

G. Check your understanding

Are these statements correct or incorrect?

- *P. vulgaris* originally came from Europe. ☐
- All snap beans are green. ☐
- Snap beans do not take very long to mature. ☐
- All beans take the same amount of time from seeding to harvest. ☐
- You harvest beans when they have grown to their full size. ☐
- Lima beans have only one form. ☐
- In the USA all peas are grown for their seeds. ☐
- A cool temperature is best for peas. ☐
- Peas are not sown until the last frost is over. ☐
- Peas are harvested in the same way for the processed and fresh markets. ☐

H. Understanding discourse

Listen to a tutor giving some advice to a student shortly before his examinations. As you listen, note down the advice the tutor gives.

UNIT 14

FARM MACHINERY (2)

A. Understanding a printed text (1)

The following text will introduce you to the topic of **the combine harvester**. Look at the way it is divided into sections and paragraphs. Pay attention to the headings and notes in the margins, and to the illustrations and captions.

Now look at these questions:

1. In the first section, does the writer describe any disadvantages of the combine harvester?
2. Is one type of combine more popular nowadays than the others?
3. What does Figure 146 illustrate?
4. What does Figure 147 illustrate?

Read the passage through and find the answers to those questions. Remember, you do not have to understand every word to answer the questions.

The Combine Harvester

Uses

1 THE combine harvester is used to harvest all types of grain and seed crops and its use has relieved farmers of much of the burden of harvest. It has not taken all the problems out of harvesting, most of which are created by weather conditions, but it has enabled farmers to rescue crops which otherwise may have been lost. Fewer acres of crops are now left to perish in the fields.

Climates

2 The combine was first developed for use in countries where the climate is more suitable for grain production and it was said that the climate in such countries as the British Isles is totally unsuitable for its use. This is not true. Combines will work in quite damp conditions, but where the conditions are not ideal, they must be used in conjunction with a grain drier where the grain can be dried to a safe moisture content for storage.

Advantages

3 Its use results in a big reduction of time and labour required to harvest a given crop because one man, or at the most two men only, are needed to operate the machine. Other labour is required to transport the grain to the farmstead, but this is not excessive. There is no stooking, carting, stacking, or any of the other work associated with harvesting when the binder is used. The combine cuts and threshes the grain in one operation.

The three main types

Combine Types

Three main types of combines are available and these are:
(a) tractor p.t.o. driven combine,
(b) tractor-drawn, engine-driven combine,
(c) self-propelled combine.

Modern trends

5 The tendency today is towards the use of the self-propelled type because of certain advantages that it has over the other two. It is more or less a self-contained unit propelled and having the threshing mechanisms powered by its own engine. This means that it is not necessary to provide a tractor to drive it and/or pull it. It is usually quite easy to manoeuvre and work in the field because all the controls used to operate

it are close to the driver; and when he becomes acquainted with them, he finds it no more difficult to drive than his tractor. Probably the worst hazard confronting the driver of a self-propelled combine is the constant cloud of dust surrounding him whilst he is working the machine. This is much worse when crop conditions are very dry. However, it is possible to have fitted to a combine, an air-conditioned cab which considerably alleviates the dust problem.

Widths of cut

6 Self-propelled combines are made in widths of cut ranging from about 1·5 m to 6 m (5-20 ft), the latter being used on the great grain-growing areas in such countries as Canada. In this country such wide cutter-bars can present problems when it is necessary to move the combine from field to field down narrow country lanes, and 365 cm (12 ft) of cut is a common size.

The other two types

7 Of the other two types of combines the main difference is the way in which they are driven but in both cases a tractor is required to haul the combine. In one, the tractor hauls the machine which is driven by an engine, whilst in the other the tractor hauls and drives the combine by the p.t.o. shaft. The width of cut of these types generally does not exceed 2 m (7 ft) but this does not mean that their output is necessarily lower than a larger-cut self-propelled combine. Combine output depends more on the capacity of the threshing and cleaning mechanism within the machine than on width of cut it will take.

Combine Layout

Working parts

8 The combine harvester can be described as a mobile threshing machine because it differs little from the stationary thresher except that a crop is fed into a combine much more evenly and continuously as it moves forward. Figure 146 shows the various working parts of a typical self-propelled combine and the flow of the crop through it. Reference should be made to this in conjunction with the following text.

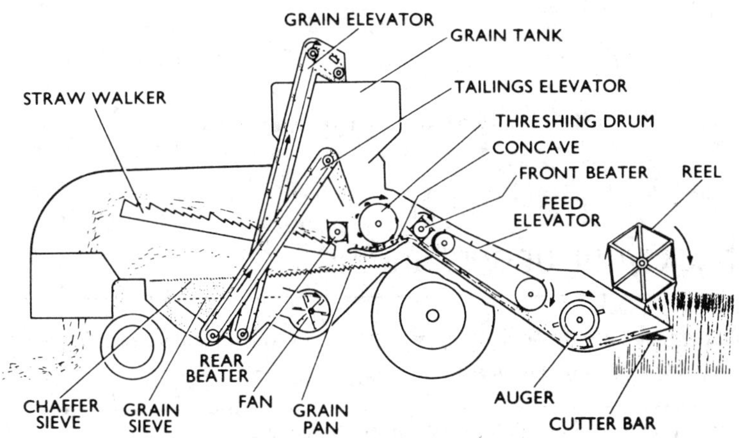

Fig. 146. A section through a combine harvester.

The cutting mechanism

The Cutter-bar

9 The cutting mechanism on a combine harvester is the same as that on a binder, consisting of a finger bar and reciprocating knife. This method of cutting crops has been used for many years now and, provided that the mechanisms are properly maintained, it is satisfactory. The mower is another machine which employs this method. The knives used on combines and binders operate at a slower speed than the mower knife because the dry straw crop is easier to cut than a swath of grass. Furthermore, it is normal practice to use a serrated edged knife on machines cutting straw crops whilst the mower uses a smooth edged knife. Very often it is found necessary to use a smooth edged knife on a combine if there is a lot of green undergrowth to be cut, but it is better to avoid cutting this undergrowth if at all possible.

10 Figure 147 shows methods used to provide knife movement. The drive to the crank pulley is usually by a V-belt and this should be kept correctly tensioned to prevent any loss of knife speed.

11 The height of the cutter-bar is adjustable vertically by mechanical or hydraulic control and it is usually arranged so that the whole cutter-bar bed and reel rises and lowers as a complete unit.

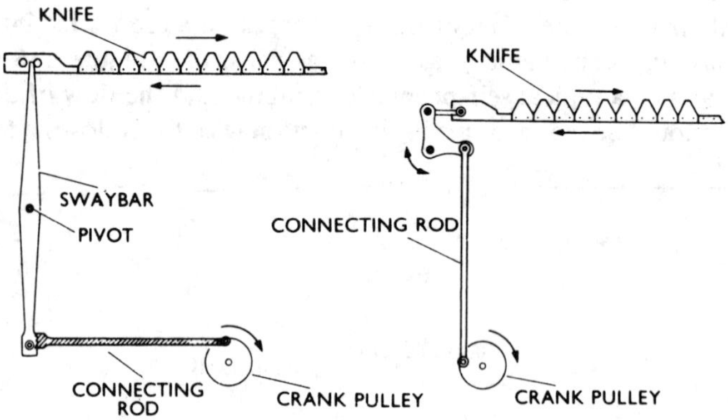

Fig. 147. Methods of providing knife movement.

Text and illustration from SHIPPEN *et al.*, *Basic Farm Machinery*, Pergamon Press

B. Check your understanding

Now read the text carefully, looking up any new items in a dictionary or reference book. Then answer the following questions:

1. Has the combine solved all the problems of harvesting?
2. Can a combine work in damp climates?
3. How must they be used in damp climates?
4. What work does a binder eliminate?
5. What danger does the driver of a self-propelled combine face?

6. What is the main difference between types (a) and (b)?
7. What does combine output depend on?
8. What does Figure 146 show in addition to the parts?
9. What do a mower and a combine have in common?
10. How do you adjust the cutter-bar?

C. Increase your vocabulary

1. Look at the first paragraph again. What words have the same meaning as:

- save
- made it possible for
- made
- something difficult or troublesome to do

2. Look at paragraph 2 again. What word has the opposite meaning to:

- without; separated from

3. Look at paragraph 3 again. Can you explain the words:

- stooking
- carting
- stacking

4. Look at paragraphs 4 and 5 again. What words have the same meaning as:

- familiar with
- move about
- danger
- facing

5. Look at paragraph 6 again and say what this word refers to:

- line 2: latter

6. Look at paragraph 7 again. What words have the same meaning as:

- be more than
- pull

7. Look at paragraph 8 again. Can you explain what is meant by:

- the flow of the crop

8. Look at paragraph 9 again. What words correspond to these definitions:

- kept in good working order
- having a toothed edge, like a saw
- ridge of wheat, grass etc, lying after it has been cut
- plants etc growing at a lower level than the crop

9. Look at paragraphs 10 and 11 again. What word has the opposite meaning to:

- horizontally

D. Check your grammar

ASKING QUESTIONS AND PARAGRAPH WRITING

You have been asked to write a description of the reel, part of a combine harvester. Your teacher has the information you need to do this. Ask questions to get the information you need. You should ask about:

1. The purpose of a reel.
2. Different types of reel.
3. Their uses.
4. Adjustments for different crop conditions.

▼

E. Understanding a lecture

1. You are now going to hear part of a lecture, divided into short sections to help you understand it. As you listen, answer the questions in your book.

Section 1

- What is the name of the part the lecturer starts with?

___

- Note down its purpose.

___

- Look at the diagram below and label it according to the information the lecturer gives you.

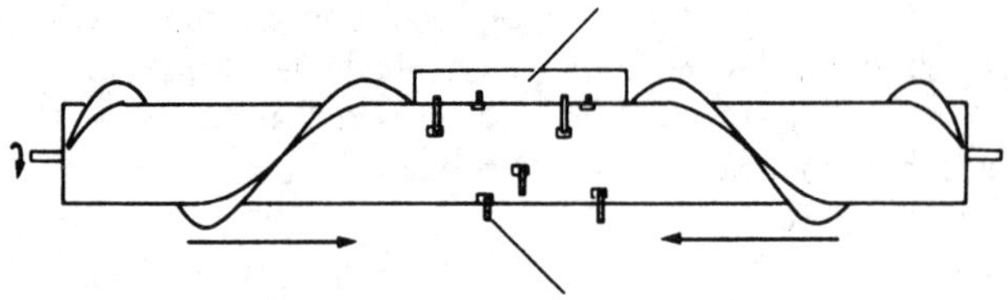

Section 2

- What part does the lecturer explain first?

___

- Note down its purpose.

___

- What is shown in the picture?

___

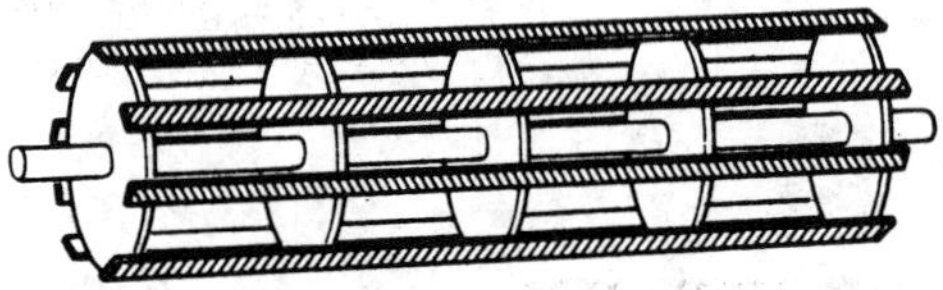

Section 3

- What is shown in the picture?

___

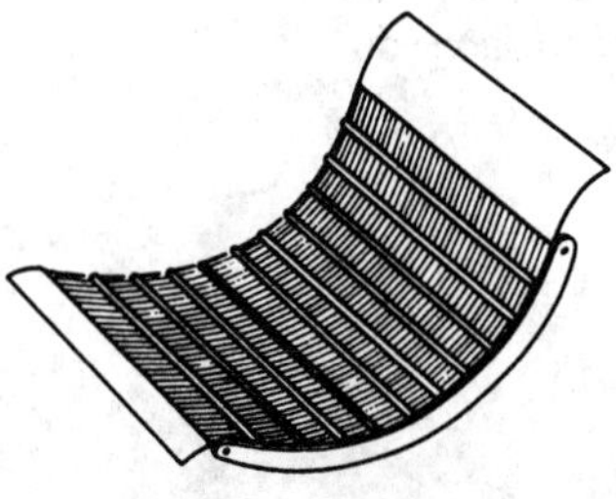

- Are these statements correct or incorrect?
  You cannot change the speed of the threshing drum. ☐
  Different crops need different degrees of threshing. ☐

Section 4

- What is another name for the rear beater?

  ___

- Note down its purpose.

  ___

- What is shown in the picture?

  ___

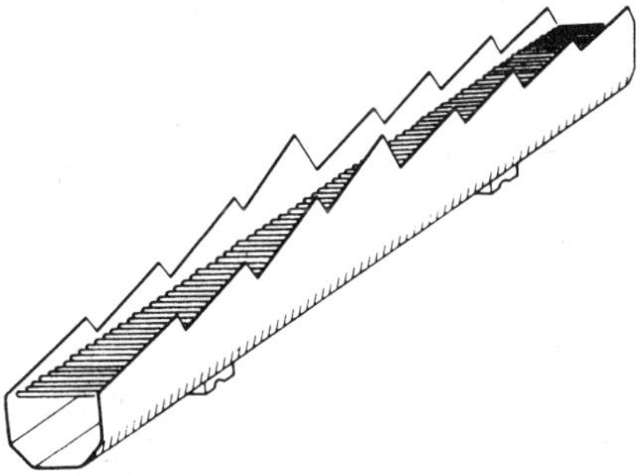

- Note down its purpose.

  ___

2. Now wind the tape back to the beginning of the lecture and listen to it again. This time, instead of answering questions, take notes. When you have listened to the whole of the lecture, you will be asked to provide a summary of it. So make sure you note down the important points.

F. Understanding a printed text (2)

The following text continues the topic of **the combine harvester**. Read it through carefully, looking up anything you do not understand.

The Grain Pan

1 The grain threshed out of the ears by the drum and concave, and the grain that comes from the straw walkers, falls on to the grain pan which is situated beneath the concave. This grain pan usually takes the form of a stepped plate occupying the full width of the cleaning unit and it also has a rearward shaking motion.

2 Whilst the grain falls through the concave and on to the pan there is also a large quantity of trash that comes through with it. Chaff, short ends of straw, weed seeds, etc., all come through mixed with the grain. This trash must be separated from the grain, so from the grain pan this mixture of trash and grain passes through a cleaning unit.

The Cleaning Unit

3 This consists of two sieves which also have a rearward shaking motion, and an air blast which can be directed up and through these sieves. The top sieve, often referred to as the chaffer sieve, extends from the grain pan and is adjustable so that the size of aperture can be increased or decreased. Grain and trash pass from the grain pan on to this sieve and the aim in setting this sieve is to set it at such a position as will allow all the grain to fall through it, whilst heavy trash is shaken rearward and out of the combine. If it is set too close, all the grain may not pass through and some may be shaken out over the rear of the combine, whilst on the other hand, if it is set too wide, excessive trash may go through it with the grain.

4 The bottom sieve, referred to as the grain sieve, is usually a sieve full of holes of a given size and is therefore not adjustable. However, this is not always the case and in some instances an adjustable sieve may be fitted. The purpose of the grain sieve is to carry out further separation of grain from trash should any trash have come through the chaffer sieve.

5 The grain sieve must be of such size as to allow the grain only to pass through and down to the grain auger. Anything larger will not pass through and will be shaken rearward to fall into the returns auger trough. Such material as broken heads will go to the returns auger.

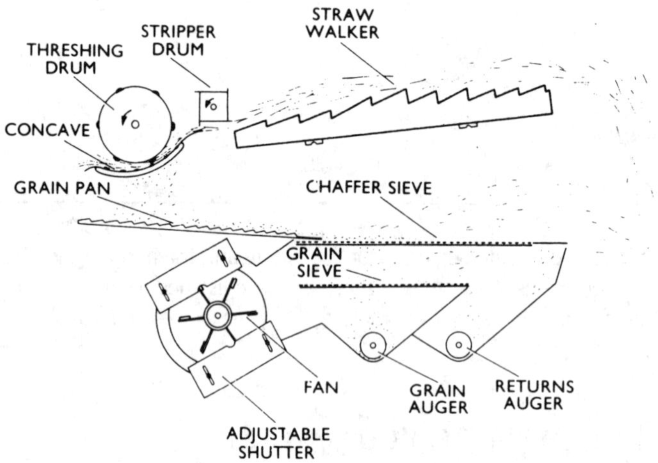

Fig. 153. The cleaning unit.

6 Whilst the grain is moving over these sieves a constant blast of air is being blown up and through them. The purpose of this is to prevent the majority of the trash from settling on the sieves. By doing this, the sieves are kept clear and the grain can fall through readily.

7 The force of the air blown through can be varied and this is necessary because the weights of different grains and seeds vary. For example, the force of air required to keep the sieves clear when combining wheat would, when combining grass seed, also blow out the seed.

8 To alter the air blast, the usual method is to reduce the inlet through which the fan draws the air. This is done by adjustable blanking-off plates. In some instances it may be possible to alter the speed at which the fan rotates by fitting a different size of pulley wheel to the fan shaft, or by a variable-speed pulley.

9 Sieves and air blast must be set in conjunction with each other to suit a particular crop and wherever possible combining should be carried out with as much air blast as possible and with the sieves set as wide as possible.

10 This will ensure that little trash is allowed to settle on the sieves and anything lighter in weight than the grain is blown out. Also, the wide sieve setting will ensure that the grain can easily pass through.

The Augers

11 The auger beneath the grain sieve transfers the grain, which at this stage should be quite free of any trash, to an elevator which elevates it to either a grain tank or to a cleaning and bagging platform.

12 The returns auger transfers any broken unthreshed heads and the like to another elevator which elevates these to a position where they can either be returned to the threshing drum for rethreshing or put back over the cleaning unit. In some cases the returns may be taken to a separate unit entirely where rethreshing is carried out if necessary and the grain is then passed back over the cleaning unit.

The Grain Tank

13 Grain from the grain elevator is put into a tank built on to the combine and which is usually of sufficient size to hold between 1500 and 1750 kg (60-70 bushels). This usually amounts to about half an hour's combining. The tank can be unloaded into trailers either whilst the combine is stationary or working. This is done by a clutch-operated auger put in and out of drive by the combine operator.

Cleaner and Bagger

14 Where a grain tank is not used, a bagging platform is provided where a man is employed sacking the grain as it is being combined. Often, before the grain is sacked, it passes through a rotating screen which takes out any weed seeds that may have escaped the cleaning unit.

Additional Attachments

15 Various devices can be fitted to combines for use in different crops and crop conditions.

16 *Grain lifters* are used when the crop has been badly laid down by the weather and they help to lift it so that it can be cut and pulled in by the reel. These lifters fit on to the cutter-bar fingers and project 38-45 cm (15-18 in. approx.) in front of the cutter-bar, and it is usual to use one per foot length of cutter-bar.

17 *Dividers* of some form or other are normally fitted to the outside of the cutter-bar and their purpose is to push through the crop ahead of the cutter-bar, separating the crop which is being cut from that which is standing. These dividers can be fitted with extensions for use in laid crops.

▼

18 *Straw spreaders* may be fitted to the rear of the combine to spread the straw across the field as it leaves the straw walkers. The spreader may take the form of a horizontal rotating disc positioned beneath the straw outlet.

19 *Chopper spreaders* may be fitted at the straw outlet, when straw leaving the combine is chopped into very short lengths and spread over the field.

20 *Windrow pick-up* attachments are used to pick up crops that have previously been cut by a mowing machine. This is often done to harvest such crops as grass seed, peas and trefoil seed. The crop is cut by a mower so that the seed can ripen in the swath and later the swath is picked up by this attachment which passes it into the combine for threshing.

21 *Grain loss-monitors* may be fitted to provide the operator with a guide to the combines performance. They consist primarily of acoustic chambers attached to the end of the straw walkers and sieves. Grain falling on these sensors has a percussion effect which creates an electrical impulse, this in turn is relayed to a control panel and displays a reading which compares with the original setting. Any increased reading shows an increased loss of grain and vice versa.

This text was taken from SHIPPEN *et al.*, *Basic Farm Machinery*, Pergamon Press

G. Check your understanding

1. Look at the first paragraph again. What words have the same meaning as:
- towards the back
- taking up
- located

2. Look at paragraph 2 again. Can you explain the words:
- trash
- chaff

3. Look at paragraph 3 again. Are these statements correct or incorrect:
- The chaffer sieve is another word for the grain pan. ☐
- Heavy trash will always pass through the chaffer sieve. ☐

4. Look at paragraph 5 again. Is this statement correct or incorrect:
- The returns auger trough does not receive grain. ☐

5. Look at paragraphs 6 to 8 again. Note down the purpose of having
- a constant air blast

6. Look at paragraphs 9 and 10 again. Is this statement correct or incorrect:
- You won't get much trash on the sieves if they are set as wide as possible with a strong air blast. ☐

7. Look at paragraphs 11 to 12 again. What word has the opposite meaning to:
- drops down

8. Look at paragraph 13 again. Can you explain the word:
- bushel

9. Look at paragraph 14 again. Is this statement correct or incorrect:

- You do not use both a grain tank and a bagging platform at the same time.

10. Look at paragraphs 15 to 21 again. Which of these additional attachments would be of most use in your country? Say which of these attachments you think would be most useful to a typical farmer and why.

H. Understanding discourse

A tutor is giving some students instructions about safety. Listen to what he says and note down his advice.

UNIT 15

FARM MANAGEMENT (2)

A. Understanding a printed text (1)

The following text will introduce you to the topic of **farm management functions**. Look at the way it is divided into sections and paragraphs. Pay attention to the headings and to the illustration and caption.

Now look at these questions:

1. What are the three main management functions?
2. According to the writer, which function is the most basic?
3. What does Figure 1–1 illustrate?
4. What is implementation?
5. How important does the writer think control is?

Read the passage through and find the answers to those questions. Remember, you do not have to understand every word to answer the questions.

MANAGEMENT FUNCTIONS

1 Earlier in this chapter, management was discussed in terms of the functions performed by managers. Included were the functions of planning, implementation, and control which many consider to be the three basic or primary functions of management.[4] The other functions can be easily included as subfunctions under one of these three. Figure 1-1 is a flowchart which summarizes the management process utilizing these three functions.

Planning

2 The planning function contains a number of steps, including the identification and definition of the problem, acquiring the initial information, identifying alternative solutions, and analyzing each alternative. Planning is the most basic management function as it means deciding on a course of action, procedure, or policy. Nothing can happen until a plan or course of action is selected. The analysis of potential plans makes use of the economic principles and budgeting techniques to be discussed in Part 2.

3 Even for the same problem, planning can be a continual process as implied by the flowchart lines in Figure 1-1. New information requires a new evaluation. The control function is a source of new information as the results of the initial plan become known. This flow of new information, including some from outside the system, requires a continual evaluation and analysis of alternatives.

Implementation

4 Once the planning process is completed, the best alternative must be selected and action taken to place the plan into operation. This requires the acquisition and organization of the necessary land, labor, machinery, livestock, and annual operating inputs. An important part of the implementation function is the financing of the necessary resources. Since implementation can take time, it must begin early enough that all required resources are available at the proper time and place.

This text was taken from KAY, *Farm Management*, 2nd edn., McGraw-Hill

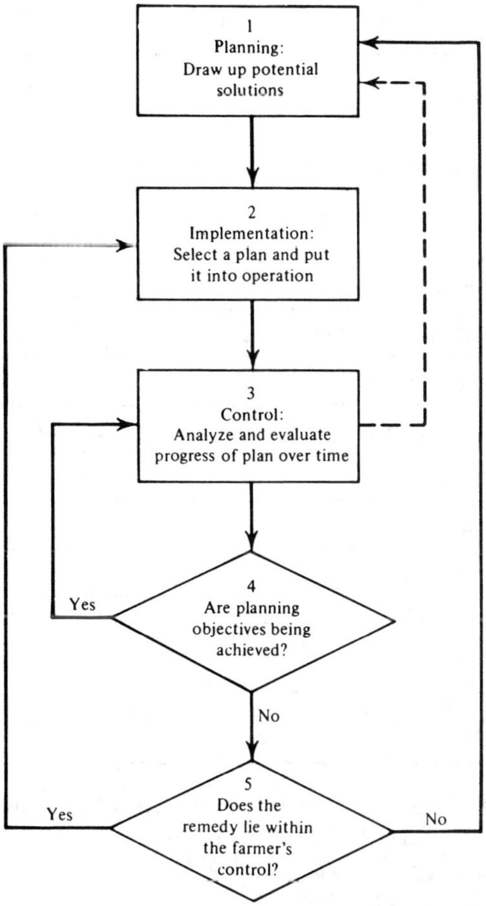

FIGURE 1-1
Management flowchart. (Adapted from C. S. Barnard and J. S. Nix, *Farm Planning and Control*, Cambridge University Press, Cambridge, England. 1973.)

Control

The control function provides for observing the results of the implemented plan to see if the specified goals and objectives are being met. Many things can cause a plan to go "off its track." Price and other changes which occur after the plan has been implemented can cause the actual results to deviate from the expected. Some deviation should be expected because of the many uncertainties connected with agricultural production, but it is necessary to identify the type and magnitude of the deviations as soon as possible. This information can be used to keep the plan and the desired results from moving outside an acceptable range.

6 Control is more than just record keeping. It requires a system for making regular checks on the plan and monitoring progress and results as measured against the established goals. The dashed line in Figure 1-1 represents the continual flow of information from the control function back to planning, an important part of the total system. Without some feedback procedure, the information obtained by the control system is of no use in making corrections to the existing plan or improving future plans. This feedback sets up a continuous cycle of planning, implementation, monitoring, and recording progress, followed by a reevaluation of the plan and the implementation procedures using the new information obtained through the control function.

7 The importance of control must be stressed continuously. A good control system requires an accurate record-keeping system and the ability to use it. Without detailed records, a business is like a ship without a compass. There is no way to know where it has been, where it is going, or how long it will take to get there.

B. Check your understanding

Now read the text carefully, looking up any new items in a dictionary or reference book. Then answer these questions:

1. What does Figure 1–1 summarise?
2. What happens if no plan is chosen?
3. What should you do with new information?
4. What do you do when you have finished the planning process?
5. What part of implementation is important?
6. What does the control function enable you to do?
7. Why should you expect deviations from the plan?
8. Is record keeping enough to ensure control?
9. Why do you need a feedback procedure?
10. What do you need in order to have a good control system?

C. Increase your vocabulary

1. Look at the first paragraph again. What words have the same meaning as:
- making use of
- believe; think

2. Look at paragraph 2 again. What nouns correspond to these verbs:
- define
- identify
- manage
- solve

3. Look at paragraph 3 again. What word has the opposite meaning to:
- at the end

4. Look at paragraph 4 again. What verbs correspond to the nouns:
- organization
- implementation
- financing
- acquisition
- function

5. Look at paragraph 5 again. What words have the same meaning as:
- watching
- move away from
- size

6. Look at paragraph 6 again. Can you explain the words:
- monitoring progress
- feedback procedure
- reevaluation

7. Look at paragraph 7 again. What word has the same meaning as:
- emphasized

D. Check your grammar

PARAGRAPH WRITING

Make complete sentences out of the following notes, putting the verbs in brackets in the right tenses. Then arrange the sentences into an introductory paragraph, and four sections headed

1. Weather
2. Land
3. Labour
4. Capital

Start with a main heading: Farm Management.

- Crops/when (grow)
- Many/with/the/him/manager/how/people (have; work)
- Weather/land/the/capital/and/labour/resources (be)
- Money/money/that/this/or/items (mean; require)
- Management/resources/a/best/proper/of/farm/its/use/the (need; achieve)
- Local/farmer/soil/his/the (must; know)
- Capital/manager/much/the/existing/how/he (must; decide; have)
- Resources/management/these/efficiently/together (mean; bring; farm)
- Weather/types/the/animal/some/of/for/unsuitable (be)
- Future/he/much/he/the/also/how/in (must work out; can obtain)
- Manager/clear/these/the/farmer/as/decisions/a/response/in/questions/to (must; make)
- Workers/the/skills/what (have)
- Available/rain/much/normally/how (be)
- Area/also/the/he/he (must measure; can use)
- Equipment/money/fertiliser/and/machines/all/tools/other (cost; buy)

E. Understanding a lecture

1. You are now going to hear part of a lecture, divided into short sections to help you understand it. As you listen, answer the questions below.

Section 1

- What is the lecturer going to talk about?
- What do these show?
- What is their main use?

Section 2

- What is the first item?
- Note down its first main use.

Section 3

- Note down its second main use.

Section 4
- What is the next item?

____

- Note down the two uses of this item.

____

Section 5
- What is the next item?

____

- Note down the two uses of this item.

____

Section 6
- What is the final item?

____

- The lecturer mentions three things this item should show. What are they?

____

2. Now wind the tape back to the beginning of the lecture and listen to it again. This time, instead of answering questions, take notes. When you have listened to the whole of the lecture, you will be asked to provide a summary of it. So make sure you note down the important points.

F. Understanding a printed text (2)

The following text continues the topic of **farm records**. Read it through carefully, looking up anything you do not understand.

CROP RECORDS

Supplemental crop records are primarily of two types, inventory records and detailed field records. Each will be discussed below.

Crop Inventory A running inventory on a monthly basis permits a manager to organize and manage crop sales, purchases, and livestock feed requirements to ensure that sufficient amounts will be available for livestock feed and for carry-over into the next year. It also permits a quick and easy method of taking a crop inventory whenever one is needed for completing a balance sheet or income statement. Several different forms can be used, and one example is shown in Table 10-1. A separate form is required for each type of grain or forage.

Each month begins with the beginning inventory, with farm production and purchases during the month added to this amount. This total is entered in the total available column. Quantities used for livestock feed, for seed, or sold during the month are subtracted from the total available to arrive at the ending inventory, and the process is repeated for the following month. It is often useful for later analyses to indicate which type of livestock consumed the feed. While some estimation will undoubtedly be necessary, a few simple notations on production and usage will provide satisfactory accuracy for this running inventory. A complete inventory at the end of each year can be used to correct any inaccuracies which have crept into the figures during the year.

Field Records Detailed field records have several important uses, including enterprise analysis on the basis of individual fields. Many managers have developed their own form for recording and maintaining this information, and an example of such a form is shown in Table 10-2. The heading contains spaces for entering the year, crop grown, field number, acres in the field, and several lines for recording weather information, effectiveness of chemicals, weed and insect problems, etc.

Information on seed variety and rates, fertilizer applications, and chemicals used can be recorded on the main part of the form. A record of chemical use is becoming very important as regulations concerning their application continue to increase. This information can also be used to plan next year's crop for each field to prevent inadvertent crop damage from herbicide carryover from one year to the next.

Table 10-2 also includes space to record the machine operations performed on the field along with the cost or hours of machinery time. If recorded in hours, this information can be used to estimate the machinery cost for the field at a later time after computing machinery cost on an hourly basis.

One objective of maintaining complete field records is to estimate profit on a field and per acre basis. Space is provided on the bottom of the form to calculate this information. The result is similar to enterprise analysis except that it is for an individual field rather than the total for all fields producing the same crop. This permits analysis on a field-by-field basis and provides more information for identifying problems and adjusting input levels based on the characteristics of an individual field.

LIVESTOCK RECORDS

Three types of livestock records will be illustrated in this section. Others are possible, including detailed input and cost records on each pen or lot of livestock, much like the field records for crop production.

Livestock Inventory A monthly livestock inventory on a form such as the one in Table 10-3 provides a convenient way to record the number of animals purchased, produced, sold, or transferred and any deaths. It also saves time when a livestock inventory is needed for completing a balance sheet or income statement or for some other purpose.

The form in Table 10-3 begins with both the number and value of the livestock on hand at the beginning of each month and ends with the same items. Additions or deletions from the inventory are recorded by number of head, including the number transferred from one class of livestock to another. The latter columns can be used when transferring replacement heifers and gilts from the inventory of market livestock to the breeding livestock inventory.

The form used to maintain the livestock inventory could be expanded to include weights and value for each of the columns. This would provide additional information but would require more space and a longer form.

Feed Records An analysis of the individual livestock enterprises or even an analysis of the profitability of the combined livestock enterprises requires feed records. Without them, it is impossible to calculate accurately profit for each livestock enterprise, returns per $100 of feed fed, or other measures of efficiency.

This text was taken from KAY, *Farm Management*, 2nd edn., McGraw-Hill

A feed record can be kept on a form such as the one shown in Table 10-4. It is designed to record feed fed to livestock on a monthly basis, but the same type of form can be used on a daily or weekly basis. Space is provided for three types of grain, three types of commercial feed, hay and silage, and two types of pasture. For feedlot livestock, the pasture columns can be eliminated to provide space for more types of the other feeds.

Special care must be taken to account for farm-raised feed fed to livestock and to price it at market value. It may be desirable to record the feed fed on both a quantity and a value basis each month. Particularly where large amounts of feed are purchased during the year at different prices, recording the feed value monthly eliminates the need to estimate an average annual price at the end of the year when calculating the total value of all feed fed.

Breeding Records The third general type of supplemental livestock record is a breeding record. Breeding records will be required for a purebred livestock operation, and to aid in the analysis and improvement of a commercial breeding herd. Several types of forms can be used for breeding records, and one type suitable for a beef cow herd is shown in Table 10-5. Most breeding records also include space to record some production and performance information to aid in culling decisions and selection of herd replacements.

The form in Table 10-5 includes spaces for recording breeding dates, sires, due dates, and actual calving dates. Limited space is available for performance data on the calf. Additional spaces could be added for other measures of cow performance such as cow weight, adjusted weaning weight, weaning index, and overall cow efficiency. This additional information can be used by the manager to form the basis for culling cows and selecting replacement heifers from the better cows.

G. Check your understanding

Using the information given in the text, and any other information source you like, complete at least one horizontal line of each of the following tables.

TABLE 10-1
MONTHLY CROP INVENTORY

Crop ____________ Year ____________

Month	Beginning inventory	Production	Purchases	Total available	Livestock feed	Farm seed	Sales	Ending inventory
January								
February								
March								
April								
May								
June								
July								
August								
September								
October								
November								
December								

TABLE 10-2
FORM FOR MAINTAINING FIELD RECORDS ON CROP PRODUCTION

Year ____________ Crop ____________ Comments: ______________________________

Field No. ____________ Acres ____________ ______________________________

____

Date	Seed		Fertilizer		Chemicals		Machinery operations	
	Variety	Rate	Analysis	Rate	Type	Rate	Type	Cost or Hours

Total Production ____________ Total Crop Value ____________ Field Profit ____________

Per Acre Yield ____________ Total Costs ____________ Profit Per Acre ____________

TABLE 10-3
MONTHLY LIVESTOCK INVENTORY

Year ____________ Type ____________ Lot or Pen No. ____________

Month	Beginning inventory		Purchased (No.)	Produced (No.)	Sold (No.)	Deaths (No.)	Transferred (No.)		Ending inventory	
	No.	Value					In	Out	No.	Value
January										
February										
March										
April										
May										
June										
July										
August										
September										
October										
November										
December										

TABLE 10-4
LIVESTOCK FEED RECORD

Kind of Livestock or Pen No. ______________________ Year __________

Month	No. of animals	Grain			Commercial feed			Forage		Pasture	
		Corn (bu.)						Hay	Silage		
January											
February											
March											
April											
May											
June											
July											
August											
September											
October											
November											
December											
Totals											
Price											
Total value											

TABLE 10-5
BEEF COW BREEDING AND PERFORMANCE RECORD

Cow number ______________ Breed ______________

Birth date or year acquired ______________ Purchased or raised ______________

Year	Date bred	Sire	Date due	Calving date	Calf information			Remarks
					Sex	Birth weight	Wean weight	

H. Understanding discourse

Listen to a tutor giving some advice on how to prepare for examinations. Note down the advice he gives.